Focus Groups

Focus Groups

Supporting Effective Product Development

Edited by
Joe Langford and Deana McDonagh

Taylor & Francis
Taylor & Francis Group
LONDON AND NEW YORK

First published 2003 by Taylor & Francis
11 New Fetter Lane, London EC4P 4EE

Simultaneously published in the USA and Canada
by Taylor and Francis Inc.,
29 West 35th Street, New York, NY 10001

Taylor & Francis is an imprint of the Taylor & Francis Group

Publisher's note:
This book was prepared from camera ready copy supplied by the authors.

Printed and bound in Great Britain by TJ International, Padstow, Cornwall.

British Library Cataloguing in Publication Data
A catalogue record for this book is available from the British Library.

Library of Congress Cataloging in Publication Data
A catalog record for this title has been requested.

ISBN 0-415-26208-9

Contents

Acknowledgements

Without the contributing authors, this book would never have happened. All of them have taken time out from their already busy work schedules to provide us with the material for the book and to deal with our follow-up enquiries. We are extremely grateful to them for their efforts. Many others have contributed – albeit unwittingly. Without the experience and knowledge gained from working with other professionals over the years, our personal inputs to the text would have been much diminished. In particular, we would like to mention colleagues at Loughborough University (particularly Anne Bruseberg) and Royal Mail. Thanks are also due to those at Taylor and Francis who have been involved with the project. These include: Tony Moore for suggesting it in the first place; the reviewers who provided helpful feedback on the original proposal; and finally, Sarah Kramer for doing her best to keep us on track and helping to get the manuscript into production.

Joe Langford
Deana McDonagh

Contributors

Christopher Baber is a senior lecturer at the University of Birmingham. He lectured on the MSc Work Design and Ergonomics course, before moving to the School of Electronic, Electrical and Computing Engineering, where he lectures on interactive systems design. His current research interests relate to human interaction with mobile and wearable technologies. Christopher has written over 40 scientific journal papers and book chapters, has written or edited four books and has made over 100 conference contributions.

Julie Barrett is a research fellow within the Research Group for Inclusive Environments at the University of Reading. She is currently working on a project to develop a 'design-for-all hospital portal' that will provide information, entertainment and communication to hospital patients. Julie has 13 years' experience working in the field of ergonomics and human factors as a researcher, lecturer and consultant, both in the UK and abroad. She has focused on the needs of older and disabled users, accessibility and the inclusive design of products. Immediately prior to joining the research group, Julie worked as a full-time web and interactive media usability consultant, specialising in accessibility. Julie holds a Bachelors degree in Psychology and a Masters degree in Ergonomics. She is currently undertaking postgraduate research concerning meeting the information needs of older people via the Internet and the design of websites that accommodate the needs of older users. She has published in professional journals and written books on products for older and disabled people.

Anne Bruseberg is a postdoctoral researcher at the HCI and multimedia group, Department of Computer Science, at the University of Bath, UK. Currently, she works on the development of principles and methods for cockpit design, by understanding the underlying factors leading to pilot errors in high workload situations. Previous research work at Loughborough University was concerned with the development of user-centred design methods – for application by industrial designers and design researchers. This involved the adaptation of user research techniques, such as focus group discussion, user product evaluation techniques, and creative user activities. After obtaining a combined degree in Human Factors/Production Engineering at Dresden University, Germany, she completed a PhD in Ergonomics at Loughborough University. Her PhD thesis investigated methods to establish operational knowledge and expertise for new system supervisory tasks in automated letter sorting offices – in order to design effective training programmes for acquiring complex process control skills.

Lee Cooper graduated from Glasgow University in 1996 with a degree in Psychology. He subsequently obtained an MSc in Work Design and Ergonomics from the University of Birmingham where he then stayed to undertake a PhD on the use of narrative in systems design, which was sponsored by NCR Knowledge Lab. He is currently employed as a human factors consultant at Hutchison 3G.

Peter Coughlan works at IDEO, a design consulting firm that helps companies become design and innovation leaders. Peter leads a group in IDEO's Palo Alto, California, location which specialises in helping organisations transform themselves using design tools and methods. Previous to this post, he led a design studio in IDEO's San Francisco location, and worked as a human factors specialist in both San Francisco and Palo Alto. Prior to working at IDEO, Peter was a member of the Design Context Lab at Nissan Design International, where he helped inform the design of next generation passenger vehicles. He also worked for Doblin Group, a strategic design planning firm in Chicago. There, he researched communication and work practices in a variety of settings to help conceive new communications products and work processes. Peter holds a BA in English Literature from Trinity College, an MA in Education from Boston University, and a PhD in Applied Linguistics from UCLA.

Helen Haines is a lecturer in Human Factors at the University of Nottingham. She was previously senior research consultant at the Institute for Occupational Ergonomics, University of Nottingham. During her time working as a consultant Helen developed an interest in participatory design and participatory ergonomics. This has resulted in a number of publications as well as forming the basis for her doctoral research.

Roger Haslam is Director of the Health and Safety Ergonomics Unit at Loughborough University (UK). His research interests include slip, trip and fall accidents; understanding and preventing other occupational and domestic accidents; ageing; health and safety; musculoskeletal disorders; and manual handling. Amongst Roger's recent work have been a number of influential government-funded projects including: *Back in Work* – the utility of total quality management (TQM) techniques as a method of tackling workplace and task-related risk factors for back pain (Department of Health/Health and Safety Executive); human factors aspects of falls amongst older people in the home (Department of Trade and Industry); health effects from use of non-keyboard computer input devices; causal factors for accidents in construction; usage and effects of medication prescribed for the treatment of anxiety and depression on work performance and safety (Health and Safety Executive). He is an editor of the scientific journals *Ergonomics* and *Applied Ergonomics*, and is currently Chair of Council of the Ergonomics Society.

Paul Herriotts leads the design ergonomics team at Land Rover, where he is involved in the development and assessment of new Land Rover vehicle designs. He has considerable experience in the field of vehicle ergonomics, having worked in the Ergonomics Department at the Motor Industry Research Association (MIRA), and also as Head of Ergonomics at BMW Group's UK design studio responsible for ergonomics input to their UK brands. He is particularly interested in vehicle design for 'third agers' and is currently undertaking a research programme with the University of Birmingham, with the aim of identifying design issues of importance to older drivers. Paul holds a Bachelors degree in Human Biology from the University of Loughborough and a Masters degree in Ergonomics from the University of Birmingham. His academic career includes spells as a researcher and lecturer at universities in mainland Europe and in the Far East. Paul passionately believes in inclusive design, and strives to ensure that vehicle design moves towards this goal. He is a regular user of focus groups as a design tool.

Wendy Ives is a trained psychologist and studied at London University. She has worked for the Research Bureau (when part of Unilever), Marplan, Pritchard Wood and was a director of Harris Research before setting up Davis Ives in 1973. Having a particular expertise in new product development, as well as development of conceptual approaches,

Wendy has pioneered new techniques for defining consumer needs and attitudes. She also has special experience on working with young people. She is actively involved with carrying out qualitative work herself. Her many years' experience coupled with her involvement and enthusiasm make her one of the most experienced researchers in this field. Wendy is a full member of the Market Research Society.

Joe Langford has nearly 20 years' experience working as a human factors/ergonomics consultant with his company Human Factors Solutions. He specialises in practical applications of human factors/ergonomics and has worked in a diverse range of areas including office equipment/systems, consumer products and postal operations. His interest in user participation in design stems from his placement as a student at the UK Home Office where he used this approach to support the design of fire service control rooms. Joe has taught ergonomics to students of Ergonomics, Industrial Design and Computer Science. He is a Council Member and Fellow of the Ergonomics Society, and is registered as a European Ergonomist.

Martin Maguire graduated with a BSc in Computer Studies and an MSc in Ergonomics. His PhD was concerned with the design of user interfaces for public information systems. Following computer graphics work at Leicester University, Martin joined HUSAT (now the Ergonomics and Safety Research Institute) at Loughborough University in 1983. His main areas of specialism include user needs analysis, interface design (particularly for non-technical people) and usability evaluation. Martin has a strong interest in the practical application of human-centred design methods and has produced a handbook for user requirements analysis as part of the EC RESPECT project. He is also actively concerned with the promotion of usability ideas and methods through networks such as Usabilitynet, the Usability Professionals Association and the British Computer Society. His consultancy work has included running focus groups and user-based evaluations for telecommunications companies to obtain public reactions to new digital TV and banking services. Currently, Martin is leading the user requirements and evaluation work in the EC IST FOREMMS and EUROCLIM projects, which are developing satellite-based monitoring systems for the forest environment and global warming in Europe.

Deana McDonagh is an Industrial Design lecturer in the Department of Design and Technology at Loughborough University. Her principle areas of research are focus group techniques; the female voice within the designing process; user involvement within the designing process; user-centred design techniques and methodologies including participative user testing; and user evaluation through the product development process. She lectures on the emotional domain within design, product semantics and design research methods at both undergraduate and postgraduate level. As an Industrial Design consultant she works closely with manufacturers of consumer products. Having gained international recognition for her design work she has also published widely in the area of design research and design education. She was Chair of the 3rd Design and Emotion Conference 2002 at Loughborough University, UK.

Elizabeth Sanders is President of SonicRim, a design research firm. She is a pioneer in the use of participatory research methods in the design of consumer, office and medical products, systems and interfaces. Liz's client relationships have included 3M, AT&T, Apple, Baxter, Coca Cola, Compaq, Hasbro, IBM, Iomega, Microsoft, Motorola, Procter & Gamble, Siemens, Steelcase, Texas Instruments, Thermos, Toro, and Xerox. Liz also teaches, conducts workshops and makes presentations about participatory research tools and methods around the world.

Aaron Sklar holds undergraduate degrees in both Psychology and Cognitive Science, as well as a graduate degree in Human Factors. As a member of IDEO's human factors group in Palo Alto, California, his role includes understanding and promoting the human experience of products, services, and environments. Through a variety of projects, ranging from implanting prosthetic heart valves to shopping in luxury retail environments, he finds the opportunity to empathise with people's expectations and needs as well as to address human capabilities and weaknesses. Aaron's favourite aspect of his work is to experience many diverse domains through getting to know people from all walks of life from all over the world.

Colin William is a psychology instructor at Columbus State Community College, and is pursuing a PhD in Cognitive and Developmental Psychology from Emory University. He has worked in user experience and human factors research since 1999, working directly with end-users of hardware and software products so that they can provide meaningful feedback to drive the product development process.

John Wilson has held teaching posts at Birmingham University and the University of California, Berkeley, and is currently Professor of Occupational Ergonomics at the University of Nottingham. Professor Wilson is Editor-in-Chief of *Applied Ergonomics*, Director of the Institute for Occupational Ergonomics and Director of the Virtual Reality Applications Research Team. He has produced over 400 publications, and more than 250 are in refereed books, journals or collections. He was awarded the Ergonomics Society Sir Frederic Bartlett Medal in 1995, for services to international ergonomics teaching and research.

CHAPTER 1

Introduction

Joe Langford and Deana McDonagh

1.1 ABOUT THIS BOOK

The fields of human factors/ergonomics and design have a common aim – to develop products and systems that successfully meet the needs of their users. A key part of achieving this aim is gaining a good understanding of the users and the ways in which they use products or systems. This includes identifying the features they value or enjoy as well as those they dislike, or which cause them problems.

There are many ways in which this knowledge can be elicited. These include user trials, questionnaire surveys and observation of people doing things in real-life situations. More recently though, there has been a growing trend to use focus group methodologies to support these more conventional methods. This is largely because the interactive and synergetic nature of group discussions (one of the central themes of focus group research) allows deeper insights into how and why people think and behave the way they do.

Focus groups are not new. Market researchers and social scientists have used them for many years and the basic principles are well defined. Designers and ergonomists have different requirements though. Whilst embracing the basic focus group concepts, they have developed techniques to help them make better use of the method to meet their specific needs.

Over the years many research papers have been published giving examples of the uses and adaptations of focus groups in human factors/ergonomics and design projects. To date though, there has not been a book which summarises this knowledge, or which provides a simple guide for ergonomists and designers who may be new to this type of research method. This book aims to fill this gap by:

- explaining what focus groups are, including the *benefits* and *limitations*;
- providing a simple *how to* guide to get the reader up and running quickly and easily;
- providing *real-life examples* of how other designers and ergonomists have used focus groups and related techniques in their work;
- providing a *toolkit* which summarises methods and techniques that can be used as part of focus group sessions.

Both human factors/ergonomics and design are wide in scope, encompassing many diverse activities. This book is aimed particularly at those involved in the following areas:

- Industrial design
- New product development
- Product and equipment design
- Information design
- User requirements analysis
- Human computer interaction
- Usability
- Work and workplace design
- Accidents, health and safety at work
- Task and systems analysis
- Training design
- Vehicle and transport design

1.2 THE STRUCTURE OF THE BOOK

After this introduction, there are three main sections to the book:

Part I. Organising and conducting a focus group: the logistics. This is a simple *how to* guide for ergonomists and designers who are new to focus groups. It provides practical advice on the key aspects of planning, recruiting participants, setting up the facilities, running sessions and analysing the results.

Part II. Focus groups in human factors/ergonomics and design: case studies. This section provides examples of how focus groups and related techniques have been used in human factors/ergonomics and design projects. The chapters are written by practitioners and researchers from Europe and North America and cover a wide range of different applications, illustrating the flexibility of the method. Where appropriate, guidelines for using focus groups in specific scenarios are given.

Part III. Focus group tools. Techniques that can be used as part of focus groups are summarised here, including creativity and thinking tools. This can be used for quick reference when planning sessions.

1.3 WHAT IS A FOCUS GROUP?

In broad terms, a focus group is a carefully planned discussion, designed to obtain the perceptions of the group members on a defined area of interest. Typically there are between five and twelve participants, the discussion being guided and facilitated by a moderator. The group members are selected on the basis of their individual characteristics as related to the topic of the session. The group-based nature of the discussions enables the participants to build on the responses and ideas of others, thus increasing the richness of the information gained.

Focus groups can be used as a self-contained research method or as part of a collection of research methods, quantitative or qualitative. Usually, they are conducted in a series, with at least three separate sessions to ensure that any patterns or trends detected are consistent. Often though, there may be many more sessions, exploring the subject area using different group compositions (e.g. with participants from different age groups or from different localities) and developing the content over time to explore different avenues. The method offers flexibility and can be used for many different purposes, including:

- obtaining general background knowledge for a new project, thus guiding the development of more detailed research, for example, the design of questionnaires;
- evaluating or gaining understanding and insight into results from other related research;
- gaining impressions and perceptions of existing or proposed services, products, programmes or organisations;
- stimulating new ideas or concepts.

The *focused group interview* was first used in the 1940s when social scientists used the technique to evaluate audience responses to radio programmes. The audience was asked to press red and green buttons whenever they heard anything that provoked a negative or positive response in them as they listened to the recording. At the end of the programme, the audience was asked to focus on the events they had recorded and to discuss the reasons for their reactions. Similar techniques were subsequently used to examine the effectiveness of wartime propaganda and training films. The history of focus groups is described in detail in Stewart and Shamdasani (1990), Morgan (1998) and Krueger (1994).

Focus group methods have been widely embraced by the marketing community, becoming one of the key tools employed by market researchers. The ability to gain in-depth understanding of users' reactions is essential to those selling or providing products or services. Early feedback to new concept ideas can prevent expensive commercial disasters. Exploration of consumers' attitudes and requirements can lead to product improvements or the development of lucrative new lines of products or services. The monitoring of consumers' reactions to advertising, publicity material and packaging helps with the selection of the best approaches and contributes to improving overall effectiveness. Greenbaum (1998) provides examples of how focus groups are used in market research and, in Chapter 3 of this book, Wendy Ives explores the market researcher's perspective in more detail.

Although the increase in the popularity of focus groups has largely been due to their use in market research, social scientists have continued to use them. Their applications range widely, for example: understanding public perceptions of mental illness; discovering reasons why individuals commit crimes; understanding participants' conceptions of what causes heart attacks. Information gained from such studies helps to inform high-level policy decisions and allocation of public resources. Barbour and Kitzinger (1999) provide more examples of the use of focus groups in social science research projects.

1.4 THE ADVANTAGES OF FOCUS GROUPS

The focus group method is a qualitative research tool. Unlike trials, experiments and physical measurement methods, qualitative research cannot provide hard quantitative data that can be subjected to statistical or numerical analysis. The main strength of qualitative research though, is its ability to gain a more in-depth understanding of the topic being explored. It can more easily deal with information and concepts that cannot be readily measured or quantified, for example, the emotional relationships between user, tasks, products and systems. Qualitative research enables researchers to discover some of the reasons why people behave the way they do, or less tangible issues such as why people feel a certain way towards a product (see Figure 1.1) or task. Examples of qualitative data collection methods are informal discussions, structured interviews and open-ended questions as part of a questionnaire survey. The focus group is a kind of interview, but instead of being conducted on a one-to-one basis it is a collective interview with a group of people.

A key benefit of focus groups is that researchers interact directly with participants. The interviewer or moderator can explore the responses given to questions or comments and thus discover more about individuals' perceptions and views. They can probe the accuracy of comments (maybe in response to non-verbal cues, such as gestures or facial expressions) and ask follow-up questions to clarify or qualify responses given. There is considerable flexibility so that, if necessary, questions can be added or modified in 'real time' to make the most of unexpected responses. Effective moderators can motivate participants to give more information and to participate fully where required. Also, face-to-face interaction enables the moderator to take account of the individual needs or characteristics of the participants and adjust their behaviour accordingly in order to encourage the information flow.

The above could be said to apply to all interview techniques, but focus groups have an additional advantage in that the group members can react to and build upon the responses and comments of others. This synergistic effect often leads to the emergence of information or the creation of ideas that would otherwise not have occurred. In addition, being part of a group can generate a feeling of security. When participants learn that they are not alone in having to deal with a sensitive or difficult issue, they often speak openly and honestly in the discussions. This can lead to the discovery of information that may have remained hidden in a one-to-one situation.

Figure 1.1 Focus groups can help to gain in-depth understanding – in this case to gain users' feedback on cycle helmets (courtesy of Royal Mail)

Focus groups are applicable to a wide range of topic areas and types of people. This includes those who might not be able to readily take part in other forms of research, for example, children or people who are not very literate. Although they are sometimes conducted in specialist facilities, focus groups can take place in a wide variety of settings and with a minimum of supporting or expensive equipment. This means that, where necessary, the focus group can be taken to the participants rather than the other way round (see Martin Maguire's example on financial services in Chapter 5). This is particularly helpful in cases where participants may find it hard to travel, for example, older or disabled people.

A great deal of information can be gained relatively efficiently and immediately with focus groups, particularly when compared with the equivalent time and effort required to gain a similar amount of information using individual interviews or postal surveys. The main reason for this is the group-based nature of the method. In just three or four sessions it is possible to gain views from a relatively large number of people – although care and time must be taken to recruit the most appropriate people using purposive sampling. The method is therefore fairly cost effective, although as Chapter 2 shows, running the sessions is the easy bit. There is a considerable amount of effort required beforehand to prepare the research and recruit participants, and then afterwards to analyse data and prepare reports. This should not be underestimated.

Both the concept of focus groups and the outcomes from them are easy for non-experts to understand. This is helpful when presenting results to others, including the sponsors of the research. The outputs are accessible because they can be presented in human terms. For example, the responses and comments of participants, whether written down as verbatim quotes, or presented in a video, are likely to be more easily understood than complex statistical analyses or tables of data. For this reason, the results of focus groups can be extremely powerful and persuasive – although this carries its own dangers, as discussed later.

1.5 THE DRAWBACKS AND LIMITATIONS

The strengths of the focus group method result from bringing together groups of individuals. Group settings, however, are more difficult to manage than one-to-one interactions and can lead to difficulties.

Discussion content. The discussion in a group interview is to a certain extent controlled by the group itself. What the group finds interesting to discuss may not be useful for the moderator's research needs. If the discussion veers off into irrelevant areas, time is lost and the moderator has to intervene to bring things back on course. This can be very time consuming if it happens often. The make up of certain groups will mean that it will happen more frequently in some sessions than others.

Dominant group members. By their very nature, group discussions mean that the participants listen to and respond to each other, and are influenced by what they see and hear. Whilst this brings benefits, it can also cause problems because the outcomes may be steered in certain directions. In some cases, there is a danger that a dominant or highly opinionated member of the group may monopolise the discussions and significantly influence the views of other participants, thus skewing the findings. There can be problems too with quieter, more reserved participants who may be hesitant to contribute their views to the discussion.

Quality of the discussion. It is impossible to predict in advance how group make-up will influence the quality of the discussion. In some cases, the discussion will be lively, energetic and revealing. In others it might be slow and lethargic, providing little information of value.

The presence of a skilled moderator can reduce the effect of all of the above problems, but they are unlikely to be completely eliminated. It is therefore important to plan sufficient separate sessions to provide confidence that the results are valid and consistent. If possible, it is also helpful to carry out research in parallel using different methods to triangulate the findings.

Focus groups generally rely on volunteer participants who are willing to put themselves out and take part in a session. For various reasons, certain types of people are unlikely to put themselves forward to take part, even with payment and other inducements. There is no guarantee, therefore, that those taking part are representative of a larger population, thus limiting the possibilities for generalisation of the outcomes. This, of course, is a drawback shared with many other research methods.

The objectivity of moderators is crucial. The moderator plays a key role in guiding and facilitating discussion and is a powerful member of the group. They can therefore bias the results by guiding the discussion in certain ways, or by knowingly or unknowingly providing cues which indicate desirable or undesirable responses. For focus groups concerning specialised areas (which is often the case in human factors/ergonomics and design) it is important that the moderator is knowledgeable enough to recognise discussion areas that could be beneficial to explore, and thus guide the discussion appropriately. Whilst it is possible to hire specialist moderators, it is likely that the best candidates for moderating such groups are the human factors/ergonomics researchers or designers themselves. It is thus important that they develop their moderation skills.

Focus groups cannot totally reflect real-life scenarios. The sessions are conducted in places and at times that are removed from the actual places and times where people have the experiences explored in the discussions. In effect, the participants have to remember or visualise experiences for the discussion, with few 'triggers' to help stimulate them. This can limit the quality of the outputs, as discussed by Peter Coughlan and Aaron Sklar in Chapter 8, so it is important to provide appropriate stimuli and prompts. In addition to this, there are sometimes marked differences between what participants *say* they do, and what they actually do in practice. The moderator may have to probe more deeply, or approach the same subject from several different angles to test that responses are consistent.

The data resulting from focus groups are largely open-ended in nature, with little structure. There is invariably a very large amount of data, mostly on video or audio tape, supported by notes and sketches. Analysis of these data to produce succinct and meaningful information can be extremely time-consuming. It is relatively easy to bias the final outcome by inclusion or exclusion of key statements, or by taking comments out of context. Care must also be taken to ensure that the focus group results are treated appropriately, and that the findings are not given more weight than is actually warranted. Verbatim comments or video recordings of live discussions may be more influential than a table of statistics, but are not necessarily more valid.

1.6 FOCUS GROUPS IN HUMAN FACTORS/ERGONOMICS AND DESIGN

Designers and ergonomists have not been slow to see the potential for using focus groups to support their work. Examples in published literature start to appear in the mid 1980s (Barkow and Rubenstein, 1986), but there is evidence that focus groups or similar methods were in use even earlier. For example, O'Brien (1981) describes the use of Shared Experience Events, which involved elements of the focus group method to facilitate the design of control rooms. At around the same time, ICE Ergonomics, a UK consumer ergonomics research and consultancy organisation, were using discussion groups for several purposes, including research into lawnmower accidents and vehicle design. Magdalen Galley, Operations Director of the Ergonomics and Safety Research Institute, Loughborough University, UK (formerly ICE Ergonomics and the HUSAT Research Institute) comments:

> As far as I can remember they were just considered to be the obvious thing to do, i.e. talk to the people who use the products, but in a structured way. We called them discussion groups and they usually had 8-12 people who were familiar with the product. We also did them with accident victims, naïve users, experienced users etc. to get different perspectives. We still use them extensively.
>
> Galley (2001)

Interestingly, she also added that very little of this work had been published. Discussions with others working in human factors/ergonomics and design have revealed that this is fairly common. Focus groups are not generally viewed as a tool for academic research. Rather, they are seen as a no-nonsense, practical solution to the problem of gaining an understanding of users and their needs, as evidenced by their inclusion in several mainstream human factors and design texts (e.g. Jordan, 1998, and Stanton, 1998). That said, more of those in the academic world are recognising the contributions that focus groups can make, and are employing them more widely – see Roger Haslam's comments in Chapter 6.

Within design and human factors/ergonomics, there is no clear and consistent terminology for describing focus groups. Published human factors/ergonomics and design literature reveals many examples of the use of the term 'focus group'. There are, however, many examples of projects that have incorporated aspects of focus group methods, but are described differently, for example, *discussion groups*, *Design Decision Groups*, *group meetings* and *user workshops*. This is partly because the term 'focus group' has only recently become widespread in the media, and many of these methods pre-date this. It is also, however, because researchers have adapted the basic focus group concepts for use in their own methods. They have added additional elements to help them achieve their objectives and, as such, consider them to be distinctly different to traditional focus groups. For the purpose of this book, however, all methods and examples that include a substantial degree of group discussion in order to achieve the objectives are considered to be relevant. We have used the generic term 'focus group' in this book purely for consistency and convenience because it is probably the most widely recognised term.

As with social sciences and market research, ergonomists and designers use focus groups for a range of different purposes, described in more detail in the following sections:

- understanding users, tasks and behaviours;
- identifying problems and establishing user and task needs;
- establishing frameworks for further research;
- evaluating existing or proposed designs;
- generating new design concepts;
- influencing and supporting decision-making.

1.6.1 Understanding Users, Tasks and Behaviours

Those working in product design or human factors/ergonomics appreciate that a good understanding of the users and their tasks or activities is vital to the successful design of products and systems. This is true whether the project objective is to design a new toaster or to define and implement preventive measures for reducing accidents.

Focus groups provide a valuable tool for gaining this type of understanding. The synergistic effect of the group discussion means that a great deal of information can be gained very quickly. Unexpected avenues can be explored if required and information is often revealed that would have remained hidden using other forms of research. Immersion in the sessions means that researchers and designers can start to build an empathic awareness and understanding of the people they are designing for.

Examples of this use of focus groups are relatively common. In this book, for example, Martin Maguire (Chapter 5) describes how focus groups were used to understand how families managed their finances as background to help create and deliver future financial services via digital TV or other household devices. Roger Haslam (Chapter 6) provides several examples where focus groups were used to gain an understanding of users and their behaviours, including research into the reasons behind accidents in the construction industry. In Chapter 4, Anne Bruseberg and Deana McDonagh explore the benefits to designers in becoming *immersed* in user research prior to concept generation.

Elsewhere, Cohen *et al.* (1996) describe how focus group data were used to help understand the human factors issues in quick service restaurant chains. Hanowski *et al.* (1998) used focus groups to investigate safety concerns amongst truck drivers. When investigating job hazards for youths working on farms, Bartels *et al.* (2000) used focus groups to identify tasks and activities that might lead to musculo-skeletal disorders and determine the participants' perceptions of the risks of such problems. Hoverstad and Kjolstad (1991) used focus groups to explore the complex relationships between working conditions and absenteeism due to illness.

Ede (2001) explains that she uses focus groups to help with the understanding of long-term issues where usability studies or surveys would not provide the information required, and where site visits are too time-consuming. She describes a focus group study where she was able to discover what system administrators' jobs over a range of companies entailed. When she started, she was given the impression that the jobs were very technical. The eventual picture resulting from the focus groups, however, was that they actually spent most of their time dealing with people.

Focus groups are a useful tool to aid awareness and understanding of issues affecting elderly people – a topic of great interest to many researchers and designers at this time. Amongst the studies concerning older members of the population are Lerner and Huey (1991), who looked at residential fire safety, and Barrett and Kirk (2000), who explored the information needs of elderly and elderly disabled people. Rogers *et al.* (1998) used focus groups to assess the constraints on daily living of healthy, active older people due to declining functionality. In Chapter 7, Julia Barrett and Paul Herriotts explain how focus groups can be tailored to work more effectively with older people.

1.6.2 Identifying Problems and Establishing User and Task Needs

In practice, the process of understanding users, tasks and behaviours as described above, will also provide researchers with a good understanding of the problems experienced. In turn, this knowledge provides a good basis for determining future user and task requirements. For example, as part of the output from their study on fire safety for older people, Lerner and Huey (1991) were able to identify key problems with existing products and technologies. They were also able to determine a set of fire-safety product needs and desirable features.

In their study on information and staff assistance needs for rail passengers, Kahmann *et al.* (1999) used discussion groups to explore the problems faced by disabled and elderly people when planning and making rail journeys. In many cases, as well as identifying problems, the focus group participants were able to provide examples of good practice. As a result, key requirements for future development of information and staff assistance on railways could then be identified.

Parsons and Wray (2000) ran a series of user workshops to investigate postal workers' perceptions of safety footwear. Areas covered included good and bad points of current footwear and possible areas for improvement. The sessions also identified the features that the users considered most important, helping to define functional specifications for future footwear. Similarly, when McDonagh-Philp and Torrens (2000) explored soldiers' views on combat gloves, focus groups provided insights into problems experienced when using the current equipment. Amongst these was the finding that the existing gloves were incompatible with basic operational tasks such as using and loading rifles! The sessions provided valuable information about soldiers' requirements and aspirations for future 'ideal' gloves (see Figure 1.2).

Figure 1.2 Focus groups were used to explore soldiers' views on combat gloves and to provide information about their requirements and aspirations for future 'ideal' gloves (McDonagh-Philp and Torrens, 2000)

Focus groups can provide a useful way to identify user requirements for information systems. Longmate *et al.* (2000) carried out a study to identify the requirements of two groups of stakeholders to inform the design of an online advice system. In addition to interviews with financial advisers, they ran discussion groups with potential users to identify and understand user experience and attitudes towards advisers, the Internet and online advice. The findings formed the basis of a set of requirements used to develop a concept interface for an online financial advice system.

When seeking cost-effective and practical methods for involving users in the design of decision support systems for use in UK agriculture, Parker (2002) adopted workshops – or focus groups – as a key method. The workshop sessions were used to identify issues including key aspects of decisions (e.g. when to act and what actions to take), data availability and types of delivery mechanisms. Techniques used included group discussions, exercises in sub-groups and ranking/rating exercises to help prioritise key areas for development. Parker argues that for small-scale developers, such as those in UK agriculture, employing such workshops is a simple and practical way of enabling user involvement.

In this book, Roger Haslam describes a study (Chapter 6) where focus groups were used to provide insights into the problems faced by those in the working population due to side effects of medication, including the impact on performance and safety in the workplace. The use of focus groups provided new data in this area because findings from laboratory studies on the use of drugs cannot easily be generalised and have little relevance to real world activities. The findings from the focus group research led to the identification of opportunities for improving the ways in which those experiencing anxiety or depression are treated.

1.6.3 Establishing Frameworks for Further Research

In common with social scientists, ergonomists and designers sometimes use focus groups as a way of initiating research in a new area. The group discussions provide a rich source of background information, helping to reveal the breadth of the topic and prioritise key areas for further investigation. For example, the findings can be used to guide the planning of further focus groups or one-to-one interviews on specific areas. Or they may indicate that setting up user trials or observational studies is a more effective way of exploring some issues and, if so, will help the researcher to design the studies so that they are effective. If the development of a questionnaire study is identified as a good way forward, the focus group findings will help guide the content of the questionnaires. They will also help ensure that the correct terminology or language style is used so that the intended target audience is comfortable with and understands the questions.

The focus groups carried out by Barrett and Kirk (2000) on information needs of elderly and elderly disabled people were set up specifically to guide more detailed research. They ran four focus groups with 20 participants to identify major themes and to aid in the creation of a questionnaire for a national survey. Similarly, in the study on video displays in public settings by Barkow and Rubenstein (1986), findings from focus groups with museum visitors were used to help formulate a questionnaire for large-scale distribution at a number of museums. In the truck driver study by Hanowski *et al.* (1998), the focus group findings were to be used to help guide the experimental design of a follow-up study where driving data was to be collected *in situ*.

In Chapter 6, Roger Haslam describes how focus groups were used in the initial phase of a study on construction site accidents to guide the main phase of the research – detailed follow-up studies of a large number of accidents.

1.6.4 Evaluating Existing or Proposed Designs

Designers and ergonomists frequently use focus groups to gain feedback on current or proposed designs for products and systems. Opinions on the validity and usefulness of focus groups for evaluation are divided though.

There is little doubt that for gaining users' *impressions* of a new product or system, focus groups can be extremely valuable (see Figure 1.3). In many cases, this level of detail is perfectly adequate, for example, when exploring emotive attributes or where the usability implications of the product are not too complex. For example, in Chapter 3 Wendy Ives describes the market researcher's approach to gaining feedback on proposed designs for a range of product types including electrical appliances, cleaning products and tableware. In these cases, the stimuli used for discussion ranged from concept boards (to illustrate key features of the design) through to working models. Similarly, at the start of a design project, Dolan *et al.* (1995) used focus groups to gain feedback on a range of both conventional and progressive designs of telephone handsets using photographs of consumers holding the handsets as stimuli. Later on in the project they ran further sessions to evaluate the design concepts produced by the human factors and industrial design specialists.

Some have questioned the suitability of focus groups for evaluation purposes, particularly with respect to more complex usability issues. Nielsen (1997) sums up the main concerns:

> To assess whether users can operate an interactive system, the only proper methodology is to sit users down, one at a time, and have them use the system.

> Because focus groups are groups, individuals rarely get the chance to explore the system on their own; instead, the moderator usually provides a product demo as the basis for discussion. Watching a demo is fundamentally different from actually using the product. There is never a question as to what to do next and you don't have to ponder the meaning of numerous screen options.

Few would disagree with this basic view – there are certainly great dangers in relying purely on focus group findings for more complex usability evaluations. There are, however, attractions in being able to carry out evaluations of designs in, or in conjunction with, focus group settings.

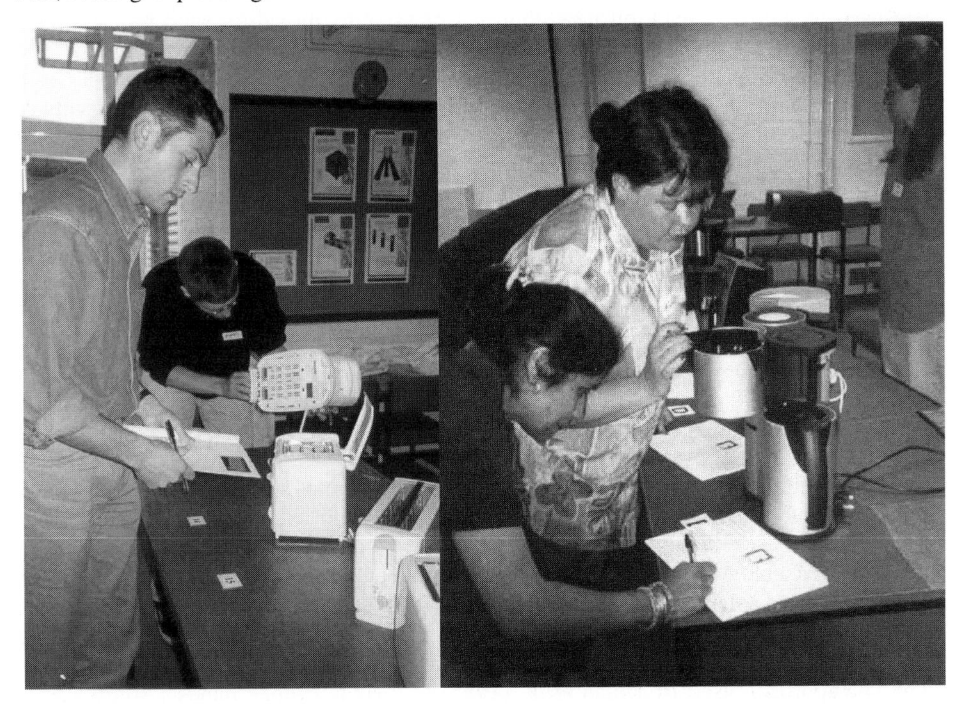

Figure 1.3 Product handling – a form of evaluation conducted within the focus group context

Caplan (1990) provides examples of this for work carried out on photocopier design. He describes some variations to the basic focus group method to yield more quantitative outputs for evaluation and to achieve valuable results with relatively small-scale efforts. One method is to use the focus group to carry out decision-making analysis (see Section 12.4.11) when comparing different design concepts for the same product. The main steps include establishing key acceptability criteria, rating the importance of each criterion and then rating the *goodness* of each concept against the criteria. Large differences between the concepts clearly point to a preferred design. Small differences require closer examination, perhaps with user trials. A second method involves participants in performance testing using prototypes prior to the focus group, then using the focus group to discuss the outcomes. The performance testing provides quantitative data and also identifies key issues for the discussion. A further benefit is that prior involvement allows the participants to become familiar with the problem area so that they can intelligently discuss it during the focus group.

O'Donnell *et al.* (1991) used an approach similar to this when they studied the use of focus groups as an evaluation technique in human computer interaction. They looked at the relationships between the findings from task-based evaluation trials and those gained from subsequent focus group sessions held some time after the trials. They found that the task-based evaluation was able to predict some, but not all, of the findings from the focus groups. Whilst the conventional measures were able to identify the main problem areas, they could not provide reasons for the errors or remedial actions. The focus groups, however, did cast some light on this – participants were able to elaborate *what was on their mind* during the task.

In Chapter 5, Martin Maguire describes several case studies where focus groups were used to gain feedback on new concepts or systems including speech-based bank machines, the use of biometrics and electronic programme guides for digital TV. Lee Cooper and Chris Baber (Chapter 9) describe how they used scenario-based discussions to help overcome the difficulties that focus group participants can have when trying to imagine how they will react to a *future* product or system. They show how they used the technique to evaluate a prototype electronic wallet. In Chapter 8, Peter Coughlan and Aaron Sklar discuss some of the problems that can occur when focus groups evaluate products and ideas in isolation from the 'real world'. They also provide some practical ways in which these problems can be reduced, thus enhancing the value of the information gained.

1.6.5 Generating New Design Concepts

The generation of new ideas is a natural step on from the evaluation of existing or proposed designs. A problem or design deficiency identified during a focus group provides a tangible baseline that can lead directly to a range of solutions. In some cases, a new or modified set of user requirements might also be identified.

Sometimes, designers and ergonomists set out with the aim of using focus group techniques to create new ideas and design concepts. In many cases, although the participants have detailed knowledge relating to their work or experiences, they are not familiar with the process of producing new ideas. The challenge here is to harness the creativity of these participants.

The approach used by Dolan *et al.* (1995) in their telephone handset study was to ask focus group participants to critique a range of handsets, and then create their own ideal designs using clay. In their study on travelling post offices (TPOs), Rainbird and Langford (1998) ran workshops with TPO staff and managers. After reviewing the existing methods and operations, the main likes, dislikes and areas for improvement were identified. The TPO users then developed their own ideas for the design of a future TPO, drawing and then presenting the ideas to the remainder of the group for discussion. Wilson (1995) describes how Design Decision Groups (DDG) can be used to enable people to participate in the analysis, redesign and evaluation of their own tasks, jobs and workplaces. The DDG process provides a structure involving group activities to analyse existing workplaces or equipment, create new designs using drawing and modelling exercises, and evaluate them with simulation trials and discussion feedback.

Design exercises within a focus group context can be extremely effective (see Figure 1.4). For example, Macbeth *et al.* (2000) carried out a study where they used focus groups to design pictorial symbols. They compared the meaningfulness of the resulting designs with sets of designs produced by subjects working as individuals and found that those produced by the focus groups were top ranked in 13 out of 17 cases. Development time for the symbols was also greatly reduced using the group method (less than one-third of the time required by the individual method).

Figure 1.4 Focus group participants taking part in a design exercise

In Chapter 10, Liz Sanders and Colin William describe some of the techniques they have developed to elicit creative thinking from participants, particularly during the early stages of design development. This includes the use of creativity-based research tools such as collaging, cognitive mapping and velcro modelling. In Chapter 11, Joe Langford, John Wilson and Helen Haines describe how group discussion methods can be used in the context of participatory design, where end-user involvement is a key part of the design development process. Examples from industrial and organisational settings are given, along with descriptions of the 'thinking tools' used to aid the process.

1.6.6 Influencing and Supporting Decision-making

A common problem faced by designers and ergonomists is how best to convey their findings to key decision-makers. Well prepared reports, models and drawings have their part to play, but are not necessarily easy for decision-makers to interpret, or enable them to appreciate the full importance of key issues. Those using focus groups have found that the evidence produced often has a strong impact, and helps them to convey their suggestions more effectively.

Caplan (1990) describes how he used videotapes from discussions to put over key points and, in some cases, involved the decision-makers in the focus group process to help them understand and accept the results. Dray and George (1992) used focus groups to identify best practices for the design of distributed information systems. They found that the outcomes were 'obvious' to the human factors professional, but the real value of the exercise was for the participants (information systems professionals and business users) to identify the issues for themselves. In Chapter 6, Roger Haslam notes that for the focus group study on the use of medication, the client specifically asked for *vignettes* giving emphasis to individual experiences to be included in the report.

Whilst not adding to the quality or content of the work, the ability to present key findings in these ways can help to achieve the objectives of the project. This fact alone may well be a compelling reason to include focus groups as part of the project approach.

1.6.7 The Focus Group as a Multi-purpose Tool

It would be highly unusual for a focus group study to be limited to just one of the above applications. There is considerable overlap between these uses, and it would actually be very hard to draw clear boundaries between some of them. For example, in focus groups aimed at evaluating existing or proposed products or services, discussion will naturally lead to suggestions for design improvement, or some completely new design ideas. Likewise, using a focus group to learn about users, tasks and behaviours will inevitably lead to findings about problems experienced and some identification of user or task needs (see Figure 1.5).

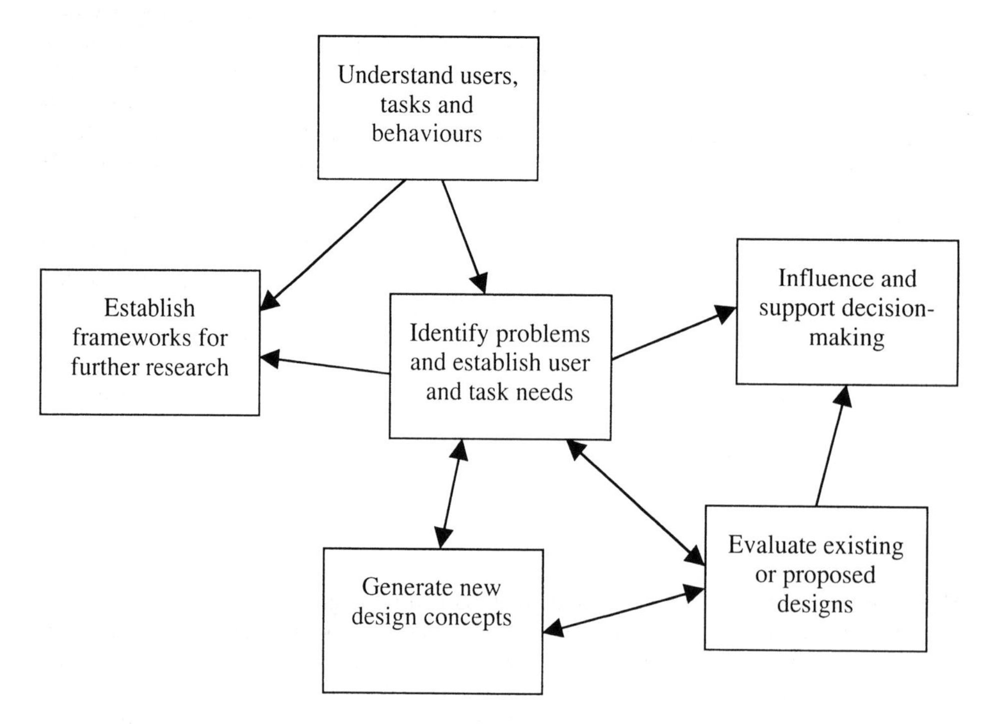

Figure 1.5 The main uses of focus groups in human factors/ergonomics and design showing the key relationships between them (Langford and McDonagh, 2002)

Blatt and Knutson (1994) provide an example of focus group sessions being designed to fulfil several purposes at once. They used focus groups in the development of an interface design guidance system (IDGS) to support system developers when creating user-system interfaces. Sessions were held with developers to solicit candid feedback on a range of issues, including: the role of the end user in the development cycle; tools currently used by developers, and their frustrations with them; developers' reactions to the IDGS concept; and, the requirements to be met if an IDGS was to be useful and adopted in software development.

The ability to use focus groups for multiple purposes is an advantage of the method. In many cases, the only way that participants can express their concerns or articulate their needs is by giving examples of the problems they experience, or describing these in the context of the tasks they perform or behaviours they adopt. Sometimes, the only way that requirements are exposed is when people generate ideas for new products or systems. The key needs or problems to be addressed will often be encoded in the drawings and sketches of their proposed solutions.

In some cases, with participatory design for example, project teams may set out with the intention of using focus group methods for different purposes at key stages throughout a project. This could include user requirements definition, concept generation, concept evaluation and deployment planning.

1.6.8 Confirming the Findings

To reduce the risk of incorrect conclusions being drawn, it is common for ergonomists and designers to carry out several different research methods in parallel, or to confirm their focus group findings with other follow-up studies (triangulation). In addition to providing confidence in the results, parallel or follow-up studies can also provide new information that would not be discoverable in a focus group environment. Where required, methods such as performance trials or analysis of records can also provide quantitative data to support conclusions.

For example, when carrying out an assessment of portable X-ray machines, Jehu and Gale (1991) used three separate approaches. These included questionnaire survey and user trials as well as participative discussion groups. In their study on slip, trip and fall accidents during delivery of mail, Bentley and Haslam (1998) conducted discussion groups with delivery employees. However, they also carried out detailed analysis of accident records, interviewed managers and safety personnel, and carried out a questionnaire survey of employees and managers. Hickling's (1984) work on safety helmets involved discussion groups, but also included a questionnaire survey and direct observation of construction workers in the field.

1.7 CONCLUSIONS

The focus group is a qualitative research method involving groups of people engaging in a planned discussion on the topic of interest. The value of focus groups is their potential to discover information that might otherwise have remained hidden with other, more formal research methods. In part, this is due to the synergistic effect of group discussions, where individuals are able to build on the ideas and responses of others. Other advantages include: applicability to a wide range of subject areas and target groups; flexibility to adapt discussions in real time; relative ease to set up and run; outputs generated in a form that is easily understood by non-experts.

There are of course some disadvantages. One frequent criticism is the potential for bias due to the group make-up, particularly the presence of dominant group members who may influence the course of the discussion. Another frequently voiced concern is that the sessions cannot truly replicate real-life scenarios, and that what people say they do can be markedly different to what they actually do.

Focus groups are popular in the fields of market research and social science, where the concept originated. Ergonomists and designers, however, have much to gain from using such techniques and, indeed, do use them regularly. Application areas include: user requirements analysis; product design; information design; human computer interaction; product usability; health and safety research; workplace, equipment, job and task design. Typical uses for focus groups in design and human factors/ergonomics include:

- understanding users, tasks and behaviours;
- identifying problems and establishing user and task needs;
- establishing frameworks for further research;
- evaluating existing or proposed designs;
- generating new design concepts;
- influencing and supporting decision-making.

Limitations of the basic discussion format have led ergonomists and designers into making adaptations to focus groups to extend their usefulness. These modifications include: *twinning* focus groups with performance testing and evaluation trials; bringing real-world experiences into the sessions, for example with 'homework' assignments; and, integration of creativity and thinking tools to aid generation of new ideas. In many cases, ergonomists and designers back up their focus group research with additional research and testing to ensure that the results are consistent and to reduce the risks of project failure.

There is no such thing as a 'perfect' research method. All have strengths and weaknesses, and focus groups are no exception. For designers and ergonomists however, they are an extremely versatile technique to add to the toolkit, providing illuminating insights into the ways people think and behave.

1.8 REFERENCES

Barbour, R.S. and Kitzinger, J., 1999, *Developing Focus Group Research. Politics, Theory and Practice*, (London: Sage Publications).

Barkow, B. and Rubenstein, R., 1986, Functional requirements for video displays in public settings. In *Proceedings of the Human Factors Association of Canada, 18th Annual Meeting, Hull*, Quebec, 27-28th September, 1985, (Rexdale, Ontario: HFAC), pp. 43-46.

Barrett, J. and Kirk, S., 2000, Running focus groups with elderly and disabled elderly participants. *Applied Ergonomics*, **31**, pp. 621-629.

Bartels, S., Niederman, B. and Waters, T.R., 2000, Job hazards for musculoskeletal disorders for youth working on farms. *Journal of Agricultural Safety and Health*, **6** (3), pp. 191-201.

Bentley, T.A. and Haslam, R.A., 1998, Slip, trip and fall accidents occurring during the delivery of mail. *Ergonomics*, **41** (12), pp. 1859-1872.

Blatt, L.A. and Knutson, J.F., 1994, Interface design guidance systems. In *Usability Inspection Methods*, Nielsen, J. and Mack, R.L. (eds), (New York: John Wiley and Sons), pp. 351-384.

Caplan, S., 1990, Using focus group methodology for ergonomic design. *Ergonomics*, **33** (5), pp. 527-533.

Cohen, S.M., Gravelle, M.D., Wilson, K.S. and Bisantz, A.M., 1996, Analysis of interview and focus group data for characterizing environments. In *Human Centered*

Technology – Key to the Future. Proceedings of the Human Factors and Ergonomics Society 40th Annual Meeting. Volume 2, 2-6 September, 1996, Philadelphia, Pennsylvania, (Santa Monica, CA: Human Factors and Ergonomics Society), pp. 957-961.

Dolan, W.R. Jr., Wiklund, M.E., Logan, R.J. and Augaitis, S., 1995, Participatory design shapes future of telephone handsets. In *Designing for the Global Village. Proceedings of the Human Factors and Ergonomics Society 39th Annual Meeting.* Volume 1, 9–13 October, 1995, San Diego, California, (Santa Monica, CA: Human Factors and Ergonomics Society), pp. 331-335.

Dray, S.M. and George, D.S., 1992, Looking back to the future: learnings about user-centered design. In *Innovations for Interactions. Proceedings of the Human Factors Society 36th Annual Meeting.* Volume 2, 12-16 October, 1992, Atlanta, Georgia, (Santa Monica, CA: Human Factors Society), pp. 872-874.

Ede, M., 2001, Focus Groups to Study Work Practice, www.useit.com/papers/meghan_focusgroup.html (accessed 31 October 2001).

Galley, M., 2001, Personal communication from the Ergonomics and Safety Research Institute, Loughborough University, Loughborough, UK.

Greenbaum, T.L, 1998, *The Handbook for Focus Group Research* (2nd edition), (Thousand Oaks: Sage Publications).

Hanowski, R.J., Wierwille, W.W., Gellatly, A.W., Knipling, R.R. and Carroll, R., 1998, Safety issues in local/short haul trucking: the drivers' perspective. In *Human-System Interaction: The Sky's No Limit. Proceedings of the Human Factors and Ergonomics Society 42nd Annual Meeting.* Volume 2, 5-9 October, 1998, Chicago, Illinois, (Santa Monica, CA: Human Factors and Ergonomics Society), pp. 1185-1189.

Hickling, E.M., 1984, *An investigation on construction sites for factors affecting the acceptability and wear of safety helmets.* Institute of Consumer Ergonomics, Loughborough, UK.

Hoverstad, T. and Kjolstad, S., 1991, Use of focus groups to study absenteeism due to illness. *Journal of Occupational Medicine*, **33** (10), pp. 1046-1050.

Jehu, F.F. and Gale, A.G., 1991, An ergonomic assessment of mobile X-ray units. In *Contemporary Ergonomics 1991*, Lovesey, E.J. (ed.), (London: Taylor and Francis), pp. 364-369.

Jordan, P.W., 1998, *An Introduction to Usability*, (London: Taylor and Francis), pp. 55-56.

Kahmann, R., Langford, J., Davis, G., Schmidt, H. and van Oel, W., 1999, COST 335 Information and Staff Training, (Amsterdam: P5 Consultants).

Krueger, R.A, 1994, *Focus Groups. A Practical Guide for Applied Research* (2nd edition), (Thousand Oaks: Sage Publications).

Langford, J. and McDonagh, D., 2002, What can focus groups offer us? In *Contemporary Ergonomics 2002*, McCabe, P.T. (ed.), (London: Taylor and Francis), pp. 502-506.

Lerner, N.D. and Huey, R.W., 1991, Residential fire safety needs of older adults. In *Visions. Proceedings of the Human Factors Society 35th Annual Meeting.* 2-6 September, 1991, San Francisco, California, (Santa Monica, CA: Human Factors Society), pp. 172-176.

Longmate, E., Lynch, P. and Baber, C., 2000, Informing the design of an online financial advice system. In *People and Computers XIV – Usability or Else! Proceedings of HCI 2000*, McDonald, S., Waern, Y. and Cockton, G. (eds), (London: Springer-Verlag), pp. 103-117.

Macbeth, S.A., Moroney, W.F. and Biers, D.W., 2000, Development and evaluation of symbols and icons: a comparison of the production and focus group method. In *Ergonomics for the New Millennium, Proceedings of the XIVth Triennial Congress of*

the International Ergonomics Association and 44th Annual Meeting of the Human Factors and Ergonomics Society. 29 July - 4 August, 2000, California, (Santa Monica, CA: Human Factors and Ergonomics Society), pp. 327-329.
McDonagh-Philp, D. and Torrens, G.E., 2000, What do British soldiers want from their gloves? In *Contemporary Ergonomics 2000*, McCabe, P.T. (ed.), (London: Taylor and Francis), pp. 349-353.
Morgan, D.L., 1998, *The Focus Group Guidebook,* (Thousand Oaks: Sage Publications).
Nielsen, J., 1997, The use and misuse of focus groups. www.useit.com/papers/focusgroups.html (accessed 12 September 1997).
O'Brien, D.D., 1981, Designing systems for new users. *Design Studies*, **2**, pp. 139-150.
O'Donnell, P.J., Scobie, G. and Baxter, I., 1991, The use of focus groups as an evaluation technique in HCI. In *People and Computers VI.* Diaper, D. and Hammond N. (eds), (Cambridge: Cambridge University Press), pp. 211–224.
Parker, C.G., 2002, Requirements analysis for decision support systems – a practical experience of a practical approach. In *Contemporary Ergonomics 2002*, McCabe, P.T. (ed.), (London: Taylor and Francis), pp. 534-539.
Parsons, C.A. and Wray, A., 2000, Specification of footwear for postal workers. In *Contemporary Ergonomics 2000*, McCabe, P.T. (ed.), (London: Taylor and Francis), pp. 344-348.
Rainbird, G. and Langford, J., 1998, Feasibility study of containerisation for travelling post office operations. In *Contemporary Ergonomics 1998*, Hanson, M.A. (ed.), (London: Taylor and Francis), pp. 381-385.
Rogers, W.A., Meyer, B., Walker, N. and Fisk, A.D., 1998, Functional limitations to daily living tasks in the aged: a focus group analysis. *Human Factors*, **40** (1), pp. 111-125.
Stanton, N. (ed.), 1998, *Human Factors in Consumer Products*, (London: Taylor and Francis).
Stewart, D.W. and Shamdasani, P.N., 1990, *Focus Groups. Theory and Practice*, (Newbury Park: Sage Publications).
Wilson, J.R., 1995, Ergonomics and participation. In *Evaluation of Human Work*, Wilson, J.R. and Corlett, E.N. (eds), (London: Taylor and Francis), pp. 1071-1096.

PART I

Organising and Conducting a Focus Group: The Logistics

CHAPTER 2

Organising and Conducting a Focus Group: The Logistics

Anne Bruseberg and Deana McDonagh

2.1 INTRODUCTION

This chapter will discuss and highlight the activities necessary to gain the most from focus group technique. Much of the procedural literature available to support the application of focus groups is tailored for social scientists and market researchers (e.g. guidebooks by Morgan (1998a, 1998b), Krueger (1988, 1998a, 1998b, 1998c), Krueger and King (1998), Greenbaum (1998)). This material is valuable and the authors have drawn on it to a large extent. However, it does not provide detailed suggestions for how focus groups can be used for applications within new product development and human factors/ergonomics. Within these areas, the focus group discussion may be combined with a range of other techniques, or may have to be adapted to suit specific purposes. The recommendations given here build on existing material, as well as experience in developing design research methods.

This chapter provides an overview of the activities that have to be considered when using focus groups. It provides a brief practical guide to support the reader in preparing, conducting and analysing data from focus group activities. The chapter gives practical advice and example illustrations on key activities involved in focus group research, including:

- the characteristics of focus groups;
- planning the research (e.g. specifying objectives, user group, costs);
- recruiting participants (e.g. contacting users, organising sessions);
- specifying the contents of the sessions (i.e. determining the structure of the sessions, the activities involved, the questions to be asked, visual aids to be used);
- setting up the sessions (e.g. room layout, payment arrangements, refreshments);
- moderating/conducting the sessions (i.e. guiding the sessions through the selected topics);
- data analysis (i.e. organising, categorising and summarising the results).

There is no such thing as a generic focus group methodology – the technique needs to be adapted to particular project aims. It is not possible to provide suggestions here for the diversity of potential applications. The most common applications may be user requirements analysis for new product development, or concept evaluation. The scope of applications, however, is very broad due to the flexibility of the technique. This chapter illustrates the main activities, with details and examples from its application to the user-centred designing process for small domestic kitchen appliances. The material provided is intended to be used as recommendations and 'food for thought', rather than a set of prescriptive rules.

2.2 WHAT ARE FOCUS GROUPS?

Focus groups are group interviews, and involve gathering together people with knowledge about a specific topic or issue (e.g. target users) for a relatively informal discussion. A chairperson or moderator promotes the discussion amongst the group, whilst carefully ensuring not to direct, but guide the group through the issues of importance to them. The *synergy* between the participants (the interaction amongst the individuals based on a mutual interest) assists in uncovering or highlighting less tangible issues. This provides an opportunity to increase understanding, awareness and empathy with group of participants interviewed (see Figure 2.1).

Figure 2.1 Focus group discussion

Focus groups provide qualitative data. The content of the discussion might take unexpected directions or open up new topics. Whilst the technique provides a high degree of flexibility in the way questions are asked, answers vary and standardisation of the data is not the focus of the research. The data provide detailed insights into people's beliefs and experiences, rather than statistically secured facts (Morgan, 1998a).

Focus groups encourage communication and provide insight into how others think and talk. They supply an efficient way of gaining an overview over various opinions at a reasonable level of detail. They provide large amounts of concentrated, well-targeted, and pre-filtered data in a short period of time. Focus groups should be used for topics that are poorly understood, because the discussion between people provides a variety of useful data. For example, focus groups provide reasons for individual opinions, and experiences. The technique is therefore ideal for early, exploratory design stages (Morgan, 1998a). For example, focus groups might uncover disregarded product functions, problems of the daily use of existing products in a range of environments, current characteristics in cultural perceptions about style and fashion, or background stories that help to visualise the users' activities and needs. The technique also offers immediate feedback for researchers, as presence during the session provides a basic idea of the main underlying messages.

Focus groups can be used to investigate complex behaviour and motivations, and to uncover subconscious notions. Through discussion, participants become more explicit about their needs. Likewise, the technique is suitable to retrieve data that are not readily formulated or knowledge not thought out in detail (Morgan, 1998a). This may be useful, as users are not always aware of all the aspects regarding the use of products, or their own preferences.

2.3 FOCUS GROUPS AS A BASE METHOD INCORPORATING A RANGE OF TECHNIQUES

Focus group activities can function both as a *technique* – enabling group discussion, in the traditional sense – as well as a *method* – being used in a broader sense to incorporate a variety of other techniques. For new product development, group discussion may provide an overall structure to research sessions, but also incorporate activities such as visual product evaluation, product handling, or drawing. Tables 2.1a and 2.1b give an overview of a range of activities and techniques that may be incorporated within a user research session. As each design brief tends to be unique in nature, the design team will need to assess the existing knowledge, experience and understanding of the team in relation to the investigation and select the most appropriate activities and techniques. Other applications may warrant a different set of techniques (Chapter 12 provides a wider range of these).

Table 2.1a Methods of feedback capture that may be used within the scope of a focus group session (for new product development)

Technique	Description	Benefits	Drawbacks
Focus group discussion	Discursive interaction between participants (e.g. a sample of users), guided by a facilitator (moderator), focusing upon particular issues	Focused data, revealing experiences and reasons for behaviour, can be elicited in a short period of time; Helps researchers to become immersed in the user's 'world of thought' and terminology; Flexible technique	Relatively high cost (e.g. recruitment administration, participant fees, data analysis); Qualitative data from conversations are difficult to formalise or use as statistically secure evidence
Questionnaires (Forms)	Retrieving feedback through use of forms with pre-determined questions (e.g. using prepared feedback boxes, rating scales and opportunities for comments)	Feedback is short and precise as comments are restricted; Tick boxes can be analysed statistically	Limited level of detail and flexibility of responses; Statistical analysis may be unsuitable due to small sample sizes; People often prefer talking to writing
Observation of activities	Watching (recording) users completing tasks or handling products	Retrieving visual information regarding the way the product is used, the features/aspects of use that are of importance to users, and emotional reactions to objects	The artefacts to be handled need to be available; Interpretation may be subjective; Users may dislike being observed
Creative participant activities	Asking participants to complete creative exercises; may be paper-based (e.g. drawing), through 3D media, or oral communication	Participants are actively involved and produce a tangible output (e.g. wish list); Users may express their wishes and aspirations more freely	Participants may be reluctant to carry out such exercises (e.g. not confident)

Table 2.1b Selected user activities that may be combined with focus groups for new product development (key: feedback capture possible through D: discussion, Q: questionnaire, O: observation, C: creative activity)

Technique	Description	Benefits	Drawbacks
Visual Evaluation	Assessing products visually based on an image only; simulating mail order/Internet purchasing conditions (see 12.4.2) *Feedback*: Q, D	Understanding the perceptions of users on product semantics, based on product appearance only	Limited information available, errors possible due to restricted image quality
Product Handling	Physical examination of products without actually using them; simulating retail showroom scenario, asking for rapid 'gut reactions' (see 12.4.3) *Feedback*: O, Q, D	Retrieving rich user feedback on products in a short amount of time; Requires limited effort	Lack of insight into problems encountered through actual product use; Lack of natural user environment (e.g. the kitchen of a user)
Mini-user trial	Users testing products in a laboratory environment by fulfilling actual tasks (see 12.4.4) *Feedback*: O, Q, D	The users gain a good idea of how the product behaves in use; Detailed issues may be revealed	Practicality and additional effort (e.g. testing kettles requires tap and sink); Safety/insurance issues (e.g. boiling water)
Product Personality Profiles	Users imagine a product as a person; Provides an insight into 'who' the user perceives to be the target user (see 12.4.6) *Feedback*: Q, D	Projective technique to retrieve hidden information; Eliciting emotional perception, user terminology, social stereotyping, and social value systems	Interpretation of the results may be complex; The participants may not like the activity and lack imagination for suitable examples
Using Mood Boards or Collages	Collection of images that represent an emotional response to the design task, the product, user etc.; Users may create the collection or it may be prepared by the design team beforehand (see 12.1.7) *Feedback*: C	Enabling users to communicate emotions non-verbally (visuals can transcend linguistic restrictions); Creating visuals that designers can use directly as inspirational material	Requires preparation or availability of suitable images; May be time-consuming; Users may be uncomfortable with the unfamiliar exercise
Brain-storming	Sharing all thoughts, ideas and comments about a particular topic – without any constraints (12.3.1) *Feedback*: Q, D, C	Novel ideas or solutions may be generated, the users' creative potential is encouraged	The user may not be accustomed to such tasks

Table 2.1b (continued)

Technique	Description	Benefits	Drawbacks
Users drawing their 'ultimate' product	Enables users to summarise and express their aspirations visually, suitable for after the discussion session (see 12.3.4) *Feedback*: C	Providing a different way of communication for the user; Provides visual stimuli for designers	Users may lack drawing skills; Users may be uncomfortable with the exercise due to lack of confidence
Creating 3D forms	Use of materials (e.g. modelling clay, cardboard) and tools by users to express their ideas and wishes (see 12.3.8) *Feedback*: C	3D representations aid the ease of communication; Perceived as easier task than drawing by users	Users may lack modelling skills; Users may be uncomfortable with the unfamiliar exercise
Nominal Group Technique	Structured group discussion technique; Involves the establishment of a consensus using a group-based categorisation and rating procedure (see 12.3.2) *Feedback*: Q, D, C	The issues are examined to a great amount of detail; There is a clear outcome of the session in the shape of a list of agreed action points or main issues	The rules of the technique are rather complex; Detailed idea examination requires additional time and moderator skill

2.4 PLANNING THE RESEARCH

It is important to determine the objectives of the research clearly at the outset. For example, the design task may vary with different types of projects, or may differ depending on the amount of information available at the beginning of the project. Likewise, design tasks requiring incremental changes to products call for a different approach to those aiming at 'blue-sky' conceptual design solutions. Moreover, realistic objectives depend on the resources available. Setting the research objectives requires establishing and specifying the following:

- purpose and required outcomes of the research, including intermediate targets;
- resources available – including personnel, skills and expertise;
- most appropriate methods and techniques to be applied;
- time scale, including internal and external deadlines;
- target participants – for example, homeowners with experience in using coffee makers.

At the outset, it has to be clear what should be achieved with the research activities and how they fit into the 'bigger picture'. It is useful to develop a time plan similar to the one shown in Figure 2.2 – specifying the different activities, time scales, and interaction with other processes (in this case, the designing process). Project requirements can vary, from conducting a single session to a whole series. This may depend on aspects such as the resources available, the type of project, and other complementary techniques to be used (e.g. home user trials). Whilst focus groups are suitable to accompany a large project through several stages, they may be applied just as an exploratory single session, or a clarifying final session (e.g. after a survey). It is often sensible to conduct at least two sessions though, as the outcomes may vary. This depends on the diversity of the participants.

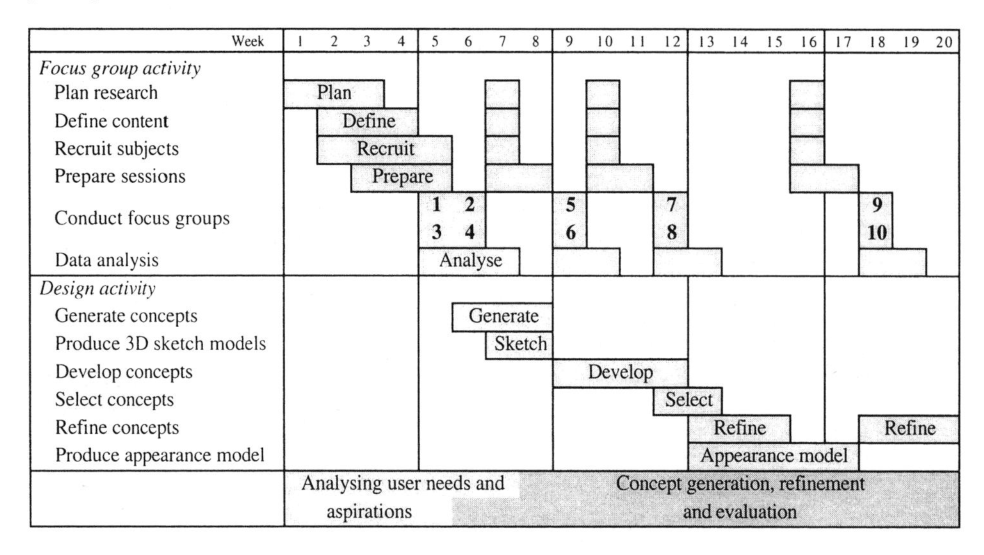

Figure 2.2 Example of a timeline for a time-constrained design project

For new product development, we envisage the use of focus groups through all the stages of the designing process – including prior to concept generation (initial requirements capture), and throughout the designing process, to evolve and evaluate concepts. Focus groups can function as a concurrent complementary technique to the designing process, enabling information to be passed directly from one stage to another (see Figure 2.3).

After a stage of broad exploration, to understand user needs and aspirations, concept generation may commence. The contents of the sessions can be adjusted flexibly according to the information requirements of the different design stages. The close interaction between user research and concept development may continue as long as required – possibly beyond the concept design stage. The process is iterative and may require going back to earlier stages before progressing with the designing process. The different activities involved in planning, defining contents, recruiting, and preparing sessions, usually overlap to some extent as they are carried out in conjunction.

Intermediate targets and deadlines are important to keep track of the progress – particularly in larger projects. Having clear objectives helps to adjust intermediate tasks in case the research reveals unexpected opportunities or problems. It is vital to identify constraints and define the variables at the outset, whilst considering the resources available (e.g. time, budget, equipment, personnel). It is useful to draw up an initial plan of which activities to include during the sessions at which stages (see Tables 2.1a and b and the tools described in Chapter 12), thus specifying the length of the sessions and the materials required.

A detailed cost plan should be designed, ensuring that all cost factors are included – taking account of aspects such as the total number of participants, fee per session, equipment and facilities needed and moderator training. Administrative costs include issues such as telephone use, producing letters, stamps, printing, photocopying, video/audio tapes, scanning, binding and batteries. Additional costs for visual stimuli and visual aids may include videos, prints, slides, renderings, drawings, models, prototypes, and product samples. The participants need to be made comfortable by providing light refreshments: the type served depends on the length of the session and the time of day/evening. There may be travel costs, required for participants and/or moderator. Room hire and other charges may need to be considered (e.g. video recorder, slide projector, multimedia projector). Paid assistance may be needed for tasks such as reception, hospitality, assistance, room presentation, administration, or preparation of materials. Moreover, indirect costs must not be overlooked – due to the time spent for preparation, analysis, and report writing.

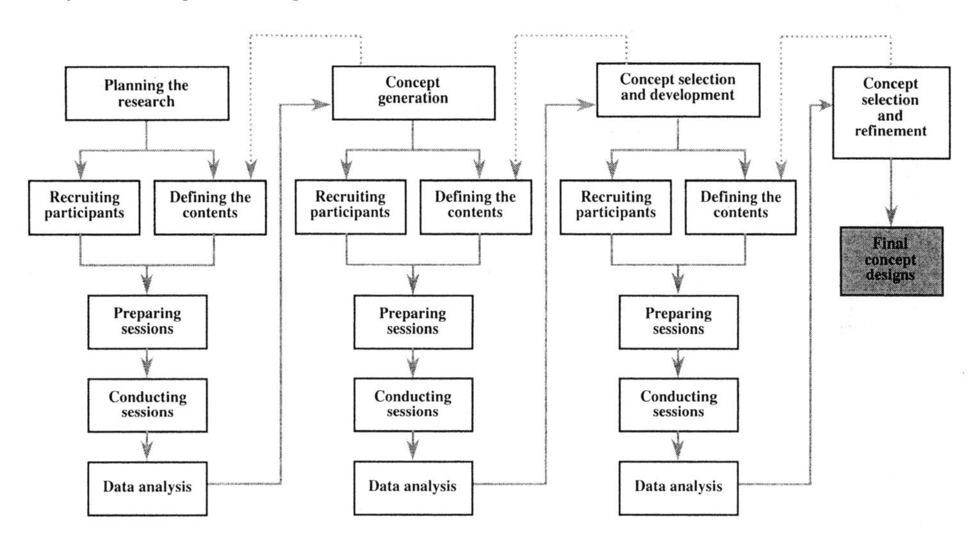

Figure 2.3 The design research process in conjunction with the use of focus groups

A professional moderator may be hired, at an additional cost. However, the moderator must be closely involved within the planning of the sessions to ensure adequate familiarity with the topic and research aims. Attendance fees for the participants and costs of hospitality might vary. The authors have paid each participant £25 per three-hour session, and have provided light refreshments. Some participants, such as expert users, or users of less mainstream products, may be more difficult to recruit and may require a higher fee. Likewise, some participants may require none (e.g. when recruiting from the client organisation).

2.5 RECRUITING PARTICIPANTS AND PLANNING SESSIONS

Prior to contacting the participants, the length of the sessions, the number of participants per group, and the main activities involved, as well as times and locations for the sessions should be known. When planning time arrangements for the sessions, a variety of factors need to be considered:

- total time that participants are required for (depending on the activities to be involved and the anticipated attention span of the participants);
- participants' fee (e.g. £30 per three-hour session);
- time of the day (e.g. working day, weekend, daytime, evening);

- intended depth of the topic (participants need to be 'warmed up' to be able to consider the topics; longer sessions may involve the participants more intensely, but they will become tired);
- number of participants attending each session.

According to the authors' experience, three hours have proved a suitable length for the sessions – combining focus group discussion with a variety of other techniques (e.g. mood boards). Providing varying activities and stimuli helps to maintain the participants' interest. Figure 2.4 illustrates how different activities can be incorporated into a three-hour research session.

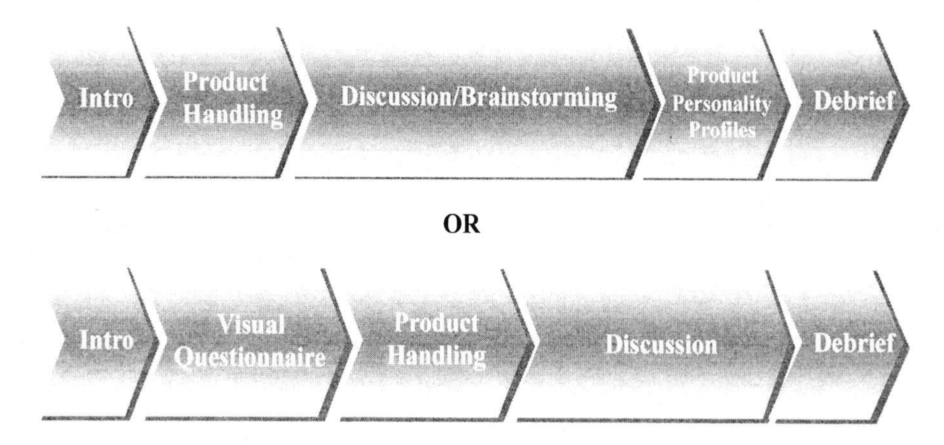

Figure 2.4 Examples of the activities that may be incorporated within a three-hour session

The contents of the session need to be tailored to the research aims at each stage of the research activities, including issues such as the scope of the project, or the profile of participants. A sufficient number of breaks of about five to ten minutes need to be provided to support participants in refocusing their concentration. Usually, the sessions need to be conducted outside of normal office hours (e.g. 9am to 5pm), unless the employer is interested in the research. It is beneficial to conduct sessions in pairs on Saturdays (one in the morning and one in the afternoon), thus ensuring people are not too tired and minimising set-up time and preparation efforts.

In the earlier stages of a project, all-male and all-female groups may be conducted on the same day with the same contents. The decision to separate genders depends on whether the issues discussed are gender-sensitive (e.g. ensuring that no participants feel the need to hold back). At later stages in the project, mixed gender groups may be held – as the issues covered tend to be less personal and recruitment becomes easier that way.

The participants should be chosen carefully through *purposive sampling* (as opposed to random sampling for surveys) – by selecting participants belonging to specific user groups (Morgan, 1998b; Erlandson *et al.*, 1993). The participants need to be reasonably knowledgeable about the topic, and should be interested in talking about it. For new product development, the profile of the target users determines the characteristics of the participants that can be selected. Some products have a wider user group, such as kettles, whilst some are more specialised, such as wheelchairs. It is important to identify from the outset if the participants should be customers or consumers, or both. The people who purchase a product (customers) are not necessarily the users (consumers). Participant criteria for group composition may include:

- occupation
- level of skill/expertise
- education
- age group
- disability
- gender
- income
- use of particular products
- home ownership

The composition of the group should not be too wide. The less rigorous the participant criteria are defined, the more variable are the outcomes. Ideally, the groups should not include too many different types of people, whilst a certain amount of diversity may be useful to encourage contrasting opinions and informative discussions. Participants need to be comfortable in talking to each other and share a similar background to encourage sharing of personal information, and more detailed insights. Different genders or age groups may offer different attitudes to products, systems or services, thus providing varying opinions.

Preferably, focus groups should consist of participants who are not too familiar with each other – especially when the session will cover sensitive issues. Over-familiarity between few members of the group may adversely affect the synergy of the group, and may make participants feel alienated (Morgan, 1998b). However, some familiarity between participants may help to 'break the ice' and encourage discussion.

These recruitment issues should be decided upon in close conjunction with the objectives of the research. In some situations (e.g. practice-based research feeding directly into an application) it may be perfectly acceptable not to conform to the general rules, whilst in others it may be vital to not violate them (e.g. large-scale academic study to investigate or establish general principles).

The time and effort required to prepare the research sessions should not be underestimated. The total time required depends on a range of factors (e.g. number of participants to be recruited, availability of resources). For example, to recruit for two focus groups involving six participants per session, it may take about two weeks (spending about two hours per day) to locate and confirm all participants' attendance, relying upon a reasonable social network. This is usually spread over a number of days. It may take more time to recruit participants that should conform to a restricted set of criteria, or experts from specific fields. Recruiting participants requires a variety of associated activities:

- planning time scales, dates and locations;
- preparing (sending out) screening questionnaires, to ensure all people meet the recruitment criteria – if a strict set of criteria has been specified;
- preparing and distributing adverts (see Figure 2.5) and invitations (see Figure 2.6);
- communicating with participants (e.g. telephone, e-mail, letters);
- composing the groups (e.g. considering participants' time preferences, ensuring a sufficient number of people in each group, dealing with late cancellations).

Traditionally, focus groups have involved around eight to ten people, though smaller groups (four to six participants) may be more suitable for design research sessions, as more time can be assigned to the individual views of participants. Some users may feel intimidated by larger groups. The presence of fewer participants will also produce a more comfortable and productive atmosphere, as a higher personal involvement may encourage people to invest more in the session. However, involving more participants provides a wider variety of opinions and ideas. The preparation effort required for small groups is very similar to that for larger groups.

The task of recruiting may either be carried out by the researchers or may be given to a recruitment agency. Suitable participants may be found by advertising in local papers, through web-based newsletters, or by word of mouth. If available, lists of suitable candidates can be used for contact through e-mail or telephone. It has to be made sure that keeping participant information complies with regulations such as the Data Protection Act (1998). For example, keeping subject details with individual questionnaire feedback must be avoided. Likewise, people's details (e.g. address) may only be kept on file beyond the project scope if the people agreed to this. Recruiting requires the preparation of confirmation letters prior to the event (see Figure 2.6), and sending out maps that help users find the venue, and a general overview of the type of activities planned for the session to inform interested candidates.

Fancy getting paid to drink coffee, eat cake and chat?

Deana McDonagh and Anne Bruseberg (Department of Design and Technology) are looking for homeowners who would like to discuss their experience of kettles, coffee makers and toasters in a lively and friendly environment.

* Each person will receive £25 for attending
* Each session will last 3 hours and involve a discussion with some handling of products
* We will provide refreshments

If you are interested, we would be delighted to hear from you via Email (a.bruseberg@lboro.ac.uk) or phone (extension 2658).

Please include the following details: (a) name, (b) age, (c) gender, and (d) occupation.

All your details will be received in confidence.

Figure 2.5 Sample call for participants (used for web-based university notice board)

Dear Jo

Thank you very much for agreeing to attend our Focus Group research session on Saturday, 21 April, 9 am to 12 noon.

The sessions will be held in the Bridgeman Centre on the Loughborough University campus (where AVS is located, close to the towers). Please find attached a map of the campus showing the location in relation to the main gate. I, or my colleague Deana, will greet you there. If the door is locked, please press the doorbell. The most convenient car park is immediately outside the AVS office.

The session will involve 6 participants. Light refreshments will be provided. Payment of £25 will be made in cash. Please bring your National Insurance number to fill in the forms on the day.

We look forward to seeing you. Please do not hesitate to contact me should you wish to discuss any details further, or require more detailed directions to the meeting point.

Thank you very much for your co-operation.

Anne

Figure 2.6 Sample invitation letter

2.6 SPECIFYING THE CONTENTS OF THE SESSIONS

2.6.1 The Moderator's Guide

A moderator's guide (see Figure 2.7 for an example) needs to be prepared in advance. This determines the contents and structure of the session, including the following:

Discussion contents: objectives of the session, activities to be included, different topics to be covered (e.g. experience of product use, imagining the future and brainstorming, questions to be asked, time allowed for each activity).

Discussion character: moderator control on the direction of the discussion, intended flow of the discussion, level of openness.

Discussion aids: visual aids and external stimuli to be used to encourage discussion.

At the outset, it is important to decide how pre-defined the structure of the sessions should be, as this influences both the formulation of questions and the moderator style to be chosen. At one end of the spectrum, the discussion can be controlled and centred on the topics provided by the moderator. The participants are kept on track to answer a list of predetermined questions within a well-defined time scale. Equally, priority may be given to the interests of the participants and the discussion may be left to flow more freely (Morgan, 1998a).

A highly structured approach may be more appropriate for the less experienced moderator, as it helps to cover situations in which there is little response from participants. However, free-flowing discussions over longer periods (e.g. 15 minutes) require very little input from the moderator, as long as they do not stray too far from the question. The degree of structure can also depend on the goals of the research. The more exploratory the study is, the fewer questions need to be predetermined, and the more scope can be given to the issues that emerge as important to the participants and relevant to the project. This means that the outcomes will be variable, comparison between groups becomes more difficult, and more groups may be required in cases where establishing a series of clear-cut messages is the main aim of the study.

Questions need to be easily understood by all the participants, using familiar terms and words. It is important to stress at the outset that the participants should use their own familiar terms. The questions need to encourage participants to express their thoughts, by preferring 'open' questions to 'closed' questions – thus avoiding questions that only leave an option to say either 'yes' or 'no'. The questions should aim to promote group discussion.

The research aims should not be set too broad, too abstract, or too demanding. Likewise, it is important to avoid influencing the discussion through evaluative terms or over-specific questions. In order to generate discussion it can be useful to prepare questions and comments that provoke a response. A well-designed moderator guide anticipates the flow of a natural conversation, linking one topic to another (Krueger 1998b).

It is important to avoid trying to be over-ambitious in the number or questions, exercises and topics to be included. Nothing is worse than having to conduct a session under time pressure. However, it is useful to have additional exercises available – either for situations in which the participants do not respond with sufficient depth to a question, or if the session progressed faster than expected. These may be left to the end, and used if needed.

Moderator's guide				
Date: 5 May	**Moderator**: Jo Bloggs **Client**: The Kettle Company **Focus Group Number**: 4	total:	**3** **0**	**hrs** **min**
Topic	Description	Aids	Duration	Start at
Pre-meeting drinks			10	9:00
Introduction	The objectives of the project; the aims/programme for the session. Practical issues: alert to video recording, confidentiality, sign consent form.	Forms in booklets	10	9:10
Visual evaluation of products (forms)	Please evaluate these products using these pictures only (like catalogue purchasing).	Product printouts, forms in booklets	30	9:20
Warm-up discussion	Please consider the most important feature when buying a new coffee maker. What makes you choose one product over another? (Ask everyone in turn if responses are slow.)		5	9:50
Understanding the contexts of product use	What are the occasions on which you use a coffee maker? Please describe them to me. Why do you own one/not own one?		10	9:55
Product handling A and form filling	Please examine these 3 products and fill in the product handling form for each of them. This is a self-contained exercise to prevent influencing others.	3 coffee makers, forms in booklets	15	10:05
Feedback discussion	Please share your view on these products.	3 coffee makers	10	10:20
Product handling B and discussion	Take product to table, ask for volunteer to assemble parts out of the box; let people handle product parts. Read out instructions from booklet. Ask for comments on criticism/positives. Retrieve views on technology, size, looks, ease of use, cleaning, coffee quality, safety, quality/durability.	New Bodum coffee maker	10	10:30
Product personality profiles	We are going to have some fun now. Please **imagine that these products were persons** with their own personality, job car, home etc. Please use the forms to fill in your responses.	Forms in booklets	15	10:40
Break			10	10:55
Presentation: The kitchen of the future	I would like to give you some ideas about **trends and future technology**. 1.Comfort kitchen – large space, including other activities, social meeting place, aesthetics important. 2. Ultra-function kitchen. 3. High-tech kitchen – intelligent, automated, speeded up.	Theme sheets with pictures, video	10	11:05
Discussion: The kitchen of the future	Now I would like to share your ideas about the **kitchen of the future** – 20 years from now. Please talk to each other. Which of the **3 trends** would you **prefer**? Do you see any **other** trends?		15	11:15
Discussion: Future devices	Stand back and think about the devices we might have in the future. How can we make things differently/better? Are there alternatives for carrying out these tasks? Why do we use the current devices? Consider their history. Would you like devices to be combined into one? Consider the clutter in the kitchen.		20	11:30
Purchasing priorities form		Forms in booklets	5	11:50
Ending session			5	11.55
				12:00

Figure 2.7 Example of a moderator's guide: data collection stage

The flow of ideas can be greatly improved by incorporating visual aids. The authors have gained positive feedback from users by continually providing idea stimulation material (e.g. various visuals, visions about future trends) as 'food for thought' throughout the sessions. External props and stimuli (i.e. models, sample products, and videos) need to be available before the session starts. Care needs to be taken not to overload the participants with visual stimuli.

2.6.2 Discussion Contents for Design Research Applications

One of the aims of carrying out user research is to enhance the lifestyle of people through providing products that respond to real user needs through more effective product development. Besides learning from existing products, user research may also aim to discover novel ideas, beyond the range of products that are currently available. Users can be engaged into creative activities to help uncover unmet user needs and to understand *ideals* – the imagination of how a perfect product could be.

It may be difficult to uncover unmet user needs. It is important that products are viewed in the context of use. Mainstream products such as kettles and toasters are products that are usually used in the domestic kitchen. Their use should not be studied in isolation. Likewise, it is much more useful to think about a device that fulfils the function of heating up water, than trying to improve on a product labelled 'kettle', as this restricts the users', as well as the designers', ability to be creative. Questioning current product roles can aid innovative problem-solving, rather than simply re-considering what already exists.

Ideals may be difficult to conceptualise and articulate for the users. In general, users are not accustomed to considering alternatives beyond what is familiar to them and may already exist. Users may be asked to consider the 'future' in order to *suspend reality*, and think more laterally and creatively. Such activities may encourage participants to disclose their wishes, aspirations and ideals more freely. Individuals require a 'warm-up' to such activities. They should not be expected to take on the role of designers. This should be made clear within the introduction of a session. Giving users creative tasks is not an attempt to find solutions, but one way to extract needs and wishes, and communicate users' suggestions to the design team.

According to the authors' experience, people enjoy talking about their routines, lifestyles, and preferences – as they feel listened to and their views are being heard. The assessment of products is also perceived as enjoyable. Initially, participants display reservations towards creative tasks, such as brainstorming and drawing, especially if they feel that their skills are inadequate. However, the results are often productive and of considerable value to the design team.

2.7 PRACTICAL CONSIDERATIONS FOR SETTING UP FOCUS GROUP SESSIONS

The preparation of focus group sessions requires a range of activities. Recruiting participants and the scheduling of sessions involves maintaining contact with participants and clarifying arrangements for the location. If the material needs to be shared between several people, or is to be analysed later in more detail, then a video recording is valuable. Participant fees and type of refreshments to be served need to be pre-determined. When not paying participants in cash, there may be a tax deduction and people should be made aware of this.

Name		Date	Signature	National Insurance number	If you do not have your National Insurance no. please give address
I, <name>	have received the payment for research participation of £25	6 Dec. 2010			
I,	have received the payment for research participation of £25	6 Dec. 2010			
I,	have received the payment for research participation of £25	6 Dec. 2010			

Figure 2.8 Sample payment form (for cash payments)

All necessary equipment and material need to be prepared in time. This includes:

- clarifying arrangements for recording (e.g. video camera) and presentation equipment (e.g. video recorder, slide projector);
- preparing forms required for payment arrangements (e.g. see Figure 2.8);
- developing questionnaires, such as consent forms (e.g. Figure 2.9), re-screening (e.g. Figure 2.10), and product evaluation questionnaires (e.g. Figure 2.11);
- providing writing supplies (a wire-bound folder for each participant is useful, containing all the forms, instructions and writing paper);
- preparing visual stimuli, such as concept models, prototypes, product samples;
- providing name cards or badges.

Subject Consent Form

Loughborough University

Department of Design and Technology
Loughborough University

I ....

consent to taking part in these **experimental sessions** to help the **evaluation** of **small domestic appliances** and elicit **user needs**.

An **explanation** of the nature and purpose of the procedure has been **given** by the experimenter.
I understand that I **may withdraw** from the experiment **at any time**, and that I am under no obligation to give reasons for withdrawal.

I understand that any **photographs** taken and **information** about myself will be treated as **confidential** by the experimenter(s).

I agree to the session being **audio taped** and **video recorded**.
I understand that what I say may be used for **research** or to compile a **report**.

Date: ....

Signed:

Signature of Researcher:

....

Figure 2.9 Sample consent form

Thank you for taking part in this study. Your views and opinions are very valuable to this project.
We would appreciate it if you could fill in the following details about yourself:

Loughborough University

1 What is your **age**? [ ] *2* **gender**? *(please tick)* male [ ] female [ ]

3 What is your present **occupation**? [ ]

4 What does your **job** involve? [ ]

5 How many paid **hours** do you work **per week**? [ ] *6* Are you a **homeowner**? *(please tick)* yes [ ] no [ ]

7 How many **people** live **in your household**? [ ] *8* How many of them are **dependent children**? [ ]

9 Did you purchase a domestic appliance within the last 6 months? *(please tick)* yes [ ] no [ ]

10 if yes, what did you buy? [ ] *11* for whom? [ ]

12 For small domestic appliances (e.g. coffee-maker), what kind of **price category** do you usually prefer purchasing at?
Do you prefer *(please tick only one)*

a) a **simple** product at the **lowest price** [ ]

b) a product with a **minimum range of functions** at a just **below average price** [ ]

c) a **good quality** product with a **few extras** at an **above average price** level [ ]

d) an expensive, **top-of-the-range**, **high quality** product that **succeeds** against most others [ ]

Figure 2.10 Re-screening form

Product handling questionnaire

Please explore the products as if you were in a retail showroom and assess the products for their ease of use.
Please feel free to pick up each product.
Use the questionnaires to fill in your responses for each product.
Please do not spend longer than 5 minutes per product.
Please feel free to ask any questions throughout this exercise.
Please do not share your opinion with other participants during the exercise (to avoid influencing others).

Product A Toaster — Product B Toaster

	very poor 1	poor 2	OK 3	good 4	very good 5	unsure 0	*Any comments?*
1 What do you think of its **visual appearance?** (e.g. style, shape, colour, texture, size)							
2 What is your perception of the **quality** (texture, solidity) of the product elements?							
3 What is your opinion about the **durability of the product?**							
4 What is your opinion about its **ease of cleaning**?							
5 What do you think about the **suitability** of this product for the following tasks:							
a) placing slices in toaster							
b) choosing required function							
c) pressing down lever							
d) monitoring progress of toasting							
e) cancelling toasting early if required							
f) lifting out toasted slices							

Figure 2.11 Product handling questionnaire

Whilst special focus group facilities (i.e. locations providing rooms with one-way mirrors, reception of participants, and catering) offer a convenient and professional environment, there may be situations where an over-formal setting may be intimidating to participants. Focus groups thrive in an open, friendly, and informal atmosphere. People tend to forget the video camera quickly, once involved in engaging activities, but one-way mirrors might be particularly unfamiliar and appear threatening. The authors recommend that observers with a high interest in the results (e.g. a designer) should be made part of the session, or may view the video later. The choice of room depends on various factors, including:

- ease of access (e.g. security arrangements, travel required, wheelchair access);
- low level of disturbance (e.g. telephones, doorbells);
- ease of setting up equipment;
- comfort for the participants (e.g. sufficient size of the room, temperature control and adequate light sources);
- facilities (e.g. toilets);
- storage location and accessibility to the visual material (product samples).

Figure 2.12a illustrates the ideal layout for a session, using a round table enabling equal face-to-face contact. Rectangular tables may be more readily available (see Figure 2.12b). Ideally, the camera/video recorder should be positioned so that facial expressions and other non-verbal communications can be recorded. It is more important to capture the participants on videotape rather than the moderator.

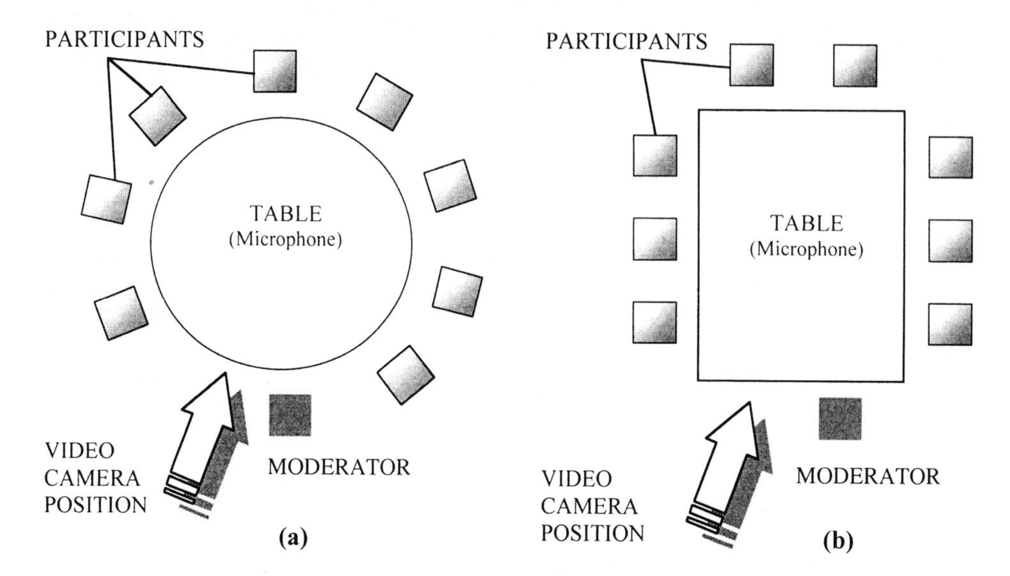

Figures 2.12 a, b Layout suggestions

The location for the refreshments and product samples within the room needs to be considered. The layout is often determined by the location of sockets, light sources, surfaces available, display boards etc. Safety is crucial when dealing with electrical equipment leads (e.g. to kettles in use, video cameras). Figure 2.13 shows an example of a room arrangement. In the situation shown, the moderator is introducing product concepts for evaluation. The moderator's assistant is making notes, but may also operate the video camera, serve drinks, take photographs, uncover or unpack product samples.

Attention to the detail is important when preparing a session. Participants are more co-operative if they feel valued and are comfortable. Providing quality food and utensils (e.g. tablecloth, ceramic mugs instead of paper ones) can make a considerable difference in creating a comfortable environment.

Figure 2.13 Layout example: the evaluation of design concepts by a female group

Key: A. Moderator B. Concept line drawings C. Video recorder
D. Moderator's assistant E. Product samples

Name cards are important as they enable people to address each other during the discussion and to feel more familiar with each other. Forenames are usually more comfortable and easier to use. This should include the moderator and all other people present (e.g. observer, assistant). Stick-on badges are very simple to use. Writing down the names also provides a suitable activity during the initial meeting period, whilst having a cup of tea and waiting for remaining participants.

2.8 CONDUCTING FOCUS GROUPS

When participants begin to arrive, it is important to:

- receive them at the entrance;
- register their names;
- show them the way to the room;
- point out facilities (e.g. toilets);
- invite them to help themselves to refreshments (or serve them);
- give participants a few minutes to make themselves comfortable and to familiarise themselves with each other.

Initially, a good introduction should prepare participants for the research session and help to make them feel welcomed. It is essential to establish basic rules for the discussion (e.g. only one person speaks at a time), to emphasise that all comments are valued, and stress that it is a non-judgmental meeting (Poulson *et al.*, 1996). Figure 2.14 shows an example of practical issues to be addressed during an introduction.

Welcome to <company name/location>. Thank you for making the time to join us this morning/afternoon. My name is <moderator's name>; I am the main researcher for today's session.

The session will last <duration> hours and involve a range of activities, including form filling and exploring products.

With your consent, we would like to videotape and photograph the session. This will help us later to analyse the material. Therefore, it is essential that we speak one at a time. Otherwise, your valuable comments may be lost.

All your feedback will be received in the strictest confidence. From now on, you will be referred to as a 'subject number', and not under your name, when analysing and presenting your feedback.

Your views and comments are immensely important to us and we need to hear all your thoughts.

We have aimed for this session to be enjoyable and informative. It is important to us that you feel comfortable. If at any time, you do not understand, or you feel uncomfortable with the session, please let us know as a matter of urgency.

The refreshments are available throughout the session and we will have dedicated breaks. The toilet facilities are to be found <here>.

Figure 2.14 Sample introduction

When moderating, it is important to find an appropriate balance between leading the session with a high level of control, and letting the discussion run on with no interruptions. It is crucial to ensure interaction amongst participants. Participants should be encouraged to avoid addressing their answers to the moderator. The moderator may literally pull away slightly from the table to encourage group interaction. It is not advisable for the moderator to take notes during the focus group discussion, unless no other recording is made, as this activity tends to stall discussion.

To formulate a new question it is useful to first introduce the new theme (e.g. 'Next, I would like you to consider the breakfast'). Then, the question needs to be stated clearly. (e.g. 'Please share with us your breakfast making procedures'). It is beneficial to underpin this with one or two additional questions or challenges (e.g. 'Do you have breakfast at all?'; 'What types of food and drink do you prepare?'). Whilst people are starting to consider their answers, it might be useful just to re-state the question to clarify (e.g. 'What do you do for your breakfast?'), and then to invite contributions and discussion.

It is important to convey only non-judgmental messages (e.g. by avoiding saying 'yes' or 'no') to continually encourage the participants, as well as to display an interest in what they are saying. It is vital to create a good atmosphere throughout the session and ensure that participants are comfortable.

Moderating also involves preventing dominant participants from having too much influence on the discussion and ensuring all participants are contributing. Extra care may be needed, as some participants may be unfamiliar with form filling, speaking publicly, or expressing their views. The moderator needs to be aware of, and responsive to, participants' skills and abilities. Table 2.2 shows some of the problems a moderator may have to deal with, and how to tackle them.

The overall structure should be regarded as quite flexible. There is often a need to extend an exercise or topic that reveals more interesting facts than expected, or takes more efforts that anticipated. On occasions, new questions arise out of the context. Then other exercises might need to be skipped or minimised.

Table 2.2 Moderating strategies to deal with problems (continued overleaf)

Problem	Measures
There are no responses to a question, or the discussion 'dries up' too quickly	There are several possible causes for lack of response: (a) *The participants are not sufficiently 'warmed up' and find the question too difficult.* Participants need some time to become accustomed with the environment and the general topic. The discussion should always start off with some 'easy' questions to which everybody should have something to say and to focus people's minds on the topic. The participants may need to think about the answers first. Participants greatly benefit from 'food for thought' (e.g. video, sample products). Example thoughts from the moderator, or a deliberately controversial opinion, are vital tools to have prepared in case the response is slow. (b) *The participants have not understood the question.* It is important to state clear questions that are not too complex (e.g. abstract thoughts, many enclosed sub-questions). It may be necessary to reformulate the question if there seems to be a misunderstanding. Highly academic terms or jargon may be particularly unsuitable, depending on the participants. (c) *The participants feel uncomfortable or intimidated by the unusual environment, or nobody wants to make a start.* It is vital that the participants are put at ease through the creation of an informal and friendly atmosphere. Humour often makes people comfortable. People need to be reassured that the moderator is really listening and all inputs are vital. Moreover, it is often useful to include exercises at the beginning, requiring input from everybody (e.g. 'How much time do you spend in your kitchen per day, doing what? Let's go round the table and start with X.'). It may be useful, on occasions, to invite the first contribution from somebody in particular, but making sure that this is a different person each time. Again, the moderator may start off the discussion with a personal example opinion. (d) *The participants are not knowledgeable about the issue or not interested in talking about it.* It might be necessary to abandon a question if all efforts fail to encourage responses. Then is important to have alternative questions available. (e) *The participants feel tired or bored.* It is vital that breaks are not being forgotten. A series of variable activities (e.g. questionnaire – video – discussion 1 – getting up to view products – discussion 2) helps to keep people interested and occupied, and are less stressful. Moreover, the enthusiasm of the moderator influences the behaviour of the participants significantly. It is valuable to introduce every task as a 'special' exercise that is going to be exciting, enjoyable, and vitally important.

Table 2.2 (continued)

Problem	Measures
Discussion flows well but is diverting from the desired topic; when to move on to the next topic?	Valuable time may be lost when letting the conversation stray to side issues. It is important that all participants have a clear idea of the direction of the question asked and are able to keep it in mind. It may be useful to provide a written memo (e.g. an overview of what was asked on a sheet in a booklet, also providing space for written notes). Discussions have their own dynamics. They can flow quite randomly depending on individual's inputs. It may be difficult to determine what is relevant, especially when the participants are the only experts. Interesting unexpected issues may be revealed when not assuming too early that the discussion content is unrelated. Often the conversation moves back on to its original path anyway. Frequent refocusing activities by the moderator may discourage input or cut off the flow. However, if the moderator feels uncomfortable with the direction of the conversation for more than 3-4 minutes, an intervention may be needed. This is done best in the shape of an actual contribution, rather than a request to come back to the original question. The moderator has to keep to the overall time scale but should allow some flexibility as to how much time to allow for each question. This needs to be handled flexibly throughout the session.
Single participants keep taking over the conversation	To reduce input from dominant personalities, the invitation to speak may deliberately be given more often to quieter members of the group. Avoiding eye contact helps to discourage too frequent input from dominant participants. They may need to be cut short occasionally, without being too aggressive. Then they often realise that they misjudged their level of input and give more space to others. It is beneficial to observe people's personalities during the initial cup of tea (before inviting people to have a seat) to identify dominant personalities. They may be invited to sit next to the moderator, thus being in a less prominent position and easier to control.
Some participants never say anything	Quiet participants require additional encouragement. This can be done occasionally by asking them directly for opinions at the beginning or the end of discussing a particular topic. This should be done infrequently though, so they do not feel like being 'picked on'. Rewarding their contributions through displaying additional attentiveness may encourage them. Controlling dominant personalities sometimes encourages more reserved participants. Once some people have said something for the first time, they then feel more confident. Hence it is useful to include exercises that require input from all at the beginning of the session. This may remain the only time they contribute, however, they have to be given their own choice. Being friendly, inviting and non-judgmental is the best strategy.

2.9 DATA ANALYSIS

Much of the analysis consists of transcribing thoughts, ideas and comments from the tapes, entering the questionnaire results (if used) into a database, and arranging the comments into suitable groups. Then general themes and categories of user needs may be identified to make the information more manageable. The data analysis can be a complex process.

For the application of new product development, it is beneficial if designers are closely involved in the data analysis process. The analysis may either be carried out directly by designers, or design researchers may examine the material and present the results to designers. Direct study of the video tapes and feedback forms by designers allows them to become immersed once more into the user reactions and ideas. People's emotional reactions, experiences and ideas may be suited best to prompt ideas in the mind of designers. By dealing with the data directly, designers are much more likely to make use of 'trigger' words and ideas, to 'spark' creative thinking. Engaging with a range of impressions may be the best way of prompting inspirations – rather than dealing with a written summarised report. A final report may be required and can function as a vehicle to communicate the findings. The actual process of transforming the data into information is when actual learning takes place, leading to increased empathy and understanding. Designers may collaborate with specialised researchers, thus saving a considerable amount of time. Having been involved in the research activities, summarised results become more meaningful to designers.

There is no need to carry out an overly extensive data analysis, such as detailed verbatim transcriptions or discourse analysis. Making notes of the essence is sufficient. Moreover, extensive comparison of data between groups or participants is not advisable unless the groups (a) have been homogeneous (regarding the types of participants as well as structure of the discussion) and (b) involved a substantial number of participants. It is beneficial to not rely on only one source of data, but to incorporate the results from the various techniques employed (i.e. conversations, average-rating scales from questionnaires used during the session, written comments) to triangulate the data.

Summarised transcriptions of conversations can be carried out with varying degrees of detail, largely depending on how much time is available. The efforts required may also vary depending on the sound quality of the recording, the clarity of expressions, and/or the variety of contrasting opinions. Experience has shown that one hour of conversation might take three to four hours to 'transcribe' and summarise. Identifying the information and making use of it goes beyond simply collecting data – it takes time and effort as well as skill. Care should be taken to reduce any bias during the analysis; ensuring equal consideration is given to the input from all participants on all aspects covered.

Figure 2.15 shows samples of how a video may be transcribed. It may be useful to determine certain symbols (e.g. abbreviations, new bullet for new thought, semicolon dividing speakers within the same line of thoughts), thus minimising efforts. The notes do not have to be in sentence format – as long as the meaning remains clear to the researcher. Likewise, certain sections may be summarised even more, thus making the message even simpler to read. When presenting the results to others, it may be useful to prepare further summarised results. Figure 2.16 shows a summary of experiences of breakfast making and eating from one group. Figure 2.17 shows a summary of the results from evaluating sample coffee-makers through questionnaires after two sessions.

Table 2.3 shows a way of categorising the results. The transcripts were analysed by establishing groups of thoughts with different ideas belonging to them. To establish an idea of importance, occurrences of mentioning a similar or the same thought were counted, thus leading to a total for each category. This may not have been influenced by the relative importance of the aspect only (e.g. some aspects may not be mentioned as they are generally 'understood'; some additional questions lead to a discussion of a topic at more depth). They only give a general idea of relative importance, but a very useful overview of the discussion contents.

What do you do in your kitchen?

- Lots of DIY (easy to clean, no other place), not eating
- When I eat on my own, I have all my meals there, coffee with friends, relaxing, has been made functional to be able to sit in there, have space
- Kitchen is entrance; same with me; also phone in the kitchen, smoke there, useful for paperwork because there is a lot of work surface
- My kitchen is a tip, need to be able to close the door
- Kitchen is not so much just for cooking anymore, now kitchen more open to visitors
- Part of my kitchen is play area for the kids – another room for them to go if other rooms are busy
- Kitchen has become more cluttered, less defined

Views on the current kitchen

- Too many gadgets, some not even used regularly, have not got enough plugs – might need to combine them?
- Mixers have improved – are more multifunctional now, has taken a lot of time to improve
- Old devices often better, new food processor not functional, difficult to clean, never use it, did it by hand in the end
- Many modern things are too complicated, experience with ultra-modern grill/toaster, couldn't work out how to do it; microwaves often too complex (defrost, weights) – couldn't work the mixer, gave it away

Appliance ideas and wishes

- Have an appliance that you can take out of the cupboard and put it on any surface and use it to cook on, put it away if you're not using it – multi-purpose, boil milk/roast on it
- I think we are snowed under with gadgets
- *There* is another gadget: to be able to run in/out, grab food, set it up in the morning, put it in, keep it cool during the day; when coming back home talk to the oven to start heating it (yes often have to wait for defrosting, then cook...), in the future we'll probably combine cooling and heating
- Machine that can learn from experience – teach it how to cook once certain food that is used regularly, to be repeated, just feed it the materials
- Just a toaster that works, toasting to the right level

Figure 2.15 Sample notes from the video (male group)

Your breakfast

- Breakfast during the week is rushed
- Some people prepare breakfast for others (kids) rather than for themselves
- Many go through a set routine and would like things to be more automated
- Some would rather like to enjoy a leisurely breakfast and most wish eating would be more of a social occasion (trend now: eat separately, food on knees in front of TV, even eat standing up)
- One member does not enjoy eating and talking and sees meals as much more functional, not as a suitable occasion for social meeting
- Maybe in the future we will have more time and revert back to social eating, taking time etc.

Figure 2.16 Sample summaries of issues from the discussion

Coffee-makers

- Although people liked product E for best appearance the winner was C – probably because it received high marks both for aesthetics and ease of use, and provided an additional function (second in appearance, balance of all aspects)
- E lost out possibly due to the poor appearance of the controls, the lack of monitoring the progress of coffee-making
- Even though D got the highest marks for ease of use, people opted for the looks, quality, durability
- Coffee-makers are more of a prestige/fun product for many – taken out on more special occasions, used less frequently
- Hence durability, timelessness, good appearance is important

Figure 2.17 Sample summaries of issues from questionnaires

Table 2.3 Extract of the category analysis for overall values regarding kitchen appliances; sorted by frequency of mentioning (right column)

REDUCING TIME	**113**
AUTOMATION: to support/help/ease; reduce boring tasks/tedious preparation/repetitive tasks, give reminders, cut down procedures (cooling-heating); don't ever hand wash – all to washer, no pots that are not dishwasher proof	32
Intelligence – dial a meal, food-sensitive, programmable, online cooking help	15
Cut out CLEANING TIME spent (e.g. washing up, tidying)	13
No time lag/INSTANT action (boiling water/drink at right temperature); speeding up basic functions (quicker oven)	12
SELF CLEANING (and drying, polishing, putting away)	12
Saving TIME important	11
Avoid extra time to do unneeded HANDGRIPS (e.g. need to use switches, decant into intelligent container, open/close toaster)	7
EFFICIENCY is vital (before/during/after use) – reduce additional handgrips (e.g. fridge to cooker), less tiring/boring things to do, quicker progress, things to work first time (do the job, no ambiguity, reliability), lack of co-ordination between appliances, easy to clean, cut down TIME/EFFORTS – most important during working week	4
Low maintenance	3
Barcodes on prepared food to be read by microwave (automatic cooking or reading out instructions); swipe in ready made meals	2
People eat pre-processed food instead of using all the gadgets they own	1
Cooking books etc. currently popular for quick cooking	1
EASE OF USE	**66**
EASE OF CLEANING is very important, part of aesthetics pleasure (e.g. surface, knobs, grooves, disassembly etc); HYGIENE (knobs dirt traps)	30
Convenience/ease of use	22
EASE OF USE/SIMPLICITY – cannot cope with too much complexity	11
Importance of accessibility – reach all	2
CO-ORDINATION of functions (walk-in gadget, powerhouse)	1
FUNCTIONALITY	**43**
DO THE JOB/FUNCTIONALITY (e.g. lack of quality for microwave)	39
Aesthetics important but kitchen is primarily FUNCTIONAL (e.g. cherished items appreciated for their function, people prioritise functionality/ease of use when considering purchase)	2
FUNCTIONALITY and the LOOK of functionality is important	1
Displays big enough	1
RESISTANCE TO AUTOMATION/DIFFERENT TECHNOLOGIES	**34**
Fear of failure of high-tech equipment/LACK OF TRUST (quality of food produced through automated devices); RELIABILITY	15
Fear of lack of control through automation/intelligence	6
Loss of skills through automation of tasks; loss of involvement, fun, tradition; makes you lazy	6
Loss of quality (e.g. taste)	3
Pre-processed food not healthy	2
Habit – don't like to miss what I'm used to (e.g. gas cooking)	1
Prefer natural things: less waste, direct access	1
SAFETY (e.g. inductive hobs, hot kettles falling); particularly when children around	**22**
SPACE	**22**
SPACE is very important, people hate CLUTTER – too many gadgets, sockets taken up, less defined	13
Devices too big – need to justify their functionality, need surface area to prepare food etc.	7
People want easy/instant ACCESS yet prefer things tidied away; everything compact, all put away but not too far (swinging units/doors)	2

It may be beneficial to collect particular phrases and quotations. The availability of a video capture card and a CD writer is recommended to be able to store passages of speech on hard disk, particularly for communicating the results to others. When presenting data from the research sessions it is advisable to 'scramble' the images of participants to protect confidentiality (e.g. by saving the file with poor video quality).

A spreadsheet such as EXCEL is a useful tool for the data analysis. It may be used to arrange quantitative questionnaire results, as well as participants' comments. Preparing a list with negative and positive adjectives for each product, based on user feedback to existing products, is a useful tool for designing (see Figure 2.18). It records participants' preferences, perceptions and language.

Product	A	B	C	D	E
Negative	Cheap Clean looking Colour universal Flimsy Little legs – different Not terribly modern Plastic – cheap Poor graphics Prefer streamlined Simple Too light Too rounded Prefer colour co-ordination Poor surface quality Large for two-slice toaster	Awkward (to use) Bulky Cheaply made Light and flimsy Wing bits at side Too complicated Too fussy Unusual/odd Ugly Too modern Quick to date Untraditional Too many lines and shapes Too many bits and pieces Prefer simple, clean lines Trying too hard (no real finesse) Pretty lines make ugly shape better Not substantial enough	Bulky base rim Cheap, fragile Fussy controls Too big/too heavy Too rounded Old-fashioned Knob protuberant	Big and bulky Bulky and clumsy Clunky Fussy Heavy Industrial looking Knobs protuberant Old-fashioned Ridges down side Too complicated Too large Too retro-looking Too many knobs and buttons Ugly controls – not balanced Things should look forward not back Commercial like catering model Too complex for simple job	Bit space age Dark edges Weird Digital switches Gimmicky Inset toast rack Light and flimsy Looks cheap Not symmetrical Too wide (long) Too serious Too big, too heavy Too many clever bits Too futuristic Overdone for toaster (not good enough) Prefer compact shape (shorter, but wide) Design for sake of being different Slick image (just a toaster) Pretentious Strange end points Over-designed
Positive	Clean Clean and smooth Clean lines Compact size Control at side Functional Looks easy to use Neat No sharp edges Nothing to offend Plain/average Practical Curves in shape/rounded contours Wavy line softens Smooth surface Unfussy Simple style/lines	Attractive Clean-looking Detailing quite good Easy to clean Fairly contemporary Friendly Good fun Light Nicer without grid	Clean Compact Curved shape Durable Fashionable Functional Good quality Sturdy base Modern colour/look Not too many dials Robust Simple clean shape Smooth and efficient-looking Blends with many colour schemes Hygienic, good surface finish Simple lines Sturdy Uncomplicated Sleek Solid, well made	Additional features All functions Clean looking durable For family use Functional Sturdy Impressive Manual controls Professional Shininess Slightly retro - streamlining Functions override shape Simple and curvy shape	Two tones Clean lines Clean styling Futuristic Good quality Impressive Modern Neat Nice balance Simple Unusual Uncomplicated Streamlined and neat Nothing sticking out to get knocked off Very eyeable

Figure 2.18 Product adjectives listing for several toasters

2.10 CONCLUSION

Focus groups provide qualitative data. The synergy between participants during discussion is a valuable tool to gain new insight. Focus groups are flexible, enabling the incorporation of a wide range of discussion topics, as well as additional techniques. They require a considerable amount of time and effort. It is important to involve the people who are going to use the results (e.g. designers) closely into planning, conducting and analysing the sessions.

Planning the research involves issues such as clarifying objectives and variables, determining time scales, and identifying potential costs. This has to be carried out in close conjunction with determining recruitment requirements and establishing the contents of the sessions. A moderators' guide is a vital tool to prepare the order of different tasks and questions, enabling a natural flow of the discussion. Questions should encourage discussion between participants. It is useful to understand people's ideals through exercises, enabling participants to suspend reality.

Preparing the sessions requires a variety of activities, including recruiting, making materials and equipment available, providing refreshments, and physically setting up the room. Conducting focus groups benefits significantly from a friendly atmosphere in which the participants feel valued. During data analysis, summaries of various levels of detail help to narrow down the richness of the data into a series of categories and overall conclusions.

The flexibility of the approach means that contents can be adjusted as required. For example, users can be involved at all stages of the concept designing process. For design-research, focus groups contribute towards evidence-based design decision-making. A large-scale study will provide in-depth information, but a 'down-sized' version may still be valuable to designers due to the qualitative character of the data.

2.11 REFERENCES

Data Protection Act, 1998, www.hmso.gov.uk/acts/acts1998/19980029.htm (accessed March 2001).

Greenbaum, T.L, 1998, *The Handbook for Focus Group Research* (2nd edition), (Thousand Oaks: Sage Publications).

Erlandson, D.A., Harris, E.L., Skipper, B.L. and Allen, S.D., 1993, *Doing Naturalistic Inquiry: A Guide to Methods*, (London: Sage Publications).

Krueger, R.A., 1988, *Focus Groups: A Practical Guide for Applied Research*, (Thousand Oaks: Sage Publications).

Krueger, R.A., 1998a, *Analysing and Reporting Focus Group Results*, (Thousand Oaks: Sage Publications).

Krueger, R.A., 1998b, *Developing Questions for Focus Groups*, (Thousand Oaks: Sage Publications).

Krueger, R.A., 1998c, *Moderating Focus Groups*, (Thousand Oaks: Sage Publications).

Krueger, R.A. and King, J.A., 1998, *Involving Community Members in Focus Groups*, (Thousand Oaks: Sage Publications).

Morgan, D.L., 1998a, *The Focus Group Guidebook*, (Thousand Oaks: Sage Publications).

Morgan, D.L., 1998b, *Planning Focus Groups*, (Thousand Oaks: Sage Publications).

Poulson, D., Ashby, M. and Richardson, S., 1996, *Userfit: a practical handbook on user-centred design for Assistive Technology*, (Brussels: ECSC-EC-EAEC).

PART II

Focus Groups in Human Factors/Ergonomics and Design: Case Studies

INTRODUCTION TO PART II

Part II provides examples of how focus groups and related techniques have been used in human factors/ergonomics and design projects. The nine chapters have been written by practitioners and researchers from Europe and North America and cover a wide range of different applications.

Overview of Part II

Focus groups have gained wide popularity in the field of market research, and market researchers are the most frequent users of the method. Market research is also closely linked with new product development. It is fitting, therefore, that the first contribution in Part II (Chapter 3) provides a market researcher's perspective on focus groups. Wendy Ives is an extremely experienced market research consultant. She describes how focus groups can help guide manufacturers and product designers in the development of new products, and provides valuable advice on how to maximise the potential of this research method.

The theme of new product development is continued in Chapter 4. Anne Bruseberg and Deana McDonagh explore designers' perceptions of focus groups, identifying the benefits and drawbacks. They present the results of three case studies carried out with practising and training designers, including a practice-based project using focus groups for user research in the design of small domestic kitchen appliances.

Martin Maguire specialises in usability and the design of human computer interfaces. In Chapter 5 he describes how focus groups can contribute to user requirements analysis. Martin includes several examples, including research into delivery of financial services via new technologies, the use of speech to interact with bank machines and the assessment of a prototype digital TV service.

Chapter 6 explores the use of focus groups in health and safety research where they have now become an established method. Roger Haslam provides three case studies; stair safety for older people, causes of accidents in the construction industry and the use of medication, including effects on work performance. Roger shows how focus groups can provide valuable insights into research questions and help to guide other research activities.

It is widely known that the world's population is ageing. Research into the design of products and services to support the older population is therefore commanding greater attention. In Chapter 7, Julia Barrett and Paul Herriotts outline the age-related declines in sensory, perceptual and communication abilities that can affect the running of focus groups with older participants. They also provide practical guidelines to help overcome these difficulties so that the method can be used more effectively.

For design consultants, gaining a deeper understanding of users' behaviour is a fundamental part of the designing process. In Chapter 8, Peter Coughlan and Aaron Sklar describe the main limitations of the 'standard' focus group method when used for this purpose. They also show how they have adapted focus groups to bring 'real world' context into the sessions, leading to a more effective design research methodology.

Lee Cooper and Chris Baber (Chapter 9) have also looked at ways of bringing context into focus groups – in this case, for helping to gather feedback on future product or service concepts. They describe a practical procedure for setting up scenario-based discussions that can help participants envision future situations and thus evaluate ideas more effectively.

In Chapter 10, Liz Sanders and Colin William describe some of the techniques they use to understand consumers and to harness the creative potential of potential end-users in the design process. These creativity-based research tools include workbooks, collages, cognitive maps and Velcro-modelling. Liz and Colin also provide practical guidance on setting up and using the tools.

User participation in design is also the subject of Chapter 11 – and, in particular, the involvement of end-users in the design of their own equipment, workplaces and jobs. Joe Langford, John Wilson and Helen Haines describe some of the tools and techniques used to help participants identify and analyse problems, and then to create and implement solutions. Two case studies are included – design of a crane control room and identification of requirements for an IT (information technology) system.

Different Backgrounds, Different Viewpoints

The contributing authors are from a diverse range of backgrounds and have employed focus groups for many different purposes. Unsurprisingly, there are differences in their perceptions of focus group methods and how they can and should be used. In some cases, the views expressed by individual authors are not in full agreement. Whilst this may initially seem unhelpful, exposure to these arguments should help the reader to more fully understand the pros and cons of focus group methods, and to make a more informed choice about how and when to use them.

CHAPTER 3

Focus Groups in Market Research

Wendy Ives

3.1 INTRODUCTION

All too often the role of market research is overlooked or pushed into the background during the design and development of a new product. This is not too surprising as market research can seem mundane when compared with the excitement of designing and developing something new. But paying insufficient attention to market research can have serious repercussions and dent the success of the new product.

There are too many examples of companies and organisations taking a new product and immediately tooling up factories ready for production without investigating the marketing aspect of the product. Often market research focus groups, carried out too late in the development process, have shown that a few product refinements would have significantly increased the marketing potential of the product. However, it is deemed too costly to incorporate them. As a result, a product is launched that does not maximise on its potential, which means that sales are curtailed.

Thus to get the most out of the new product being designed, it is vital that designers take on board the need for market research and the role that they must play, in this context, if the full value of their products are to be realised.

3.2 THE CHANGING CLIMATE

The popularity and the perceived value and relevance of focus groups in market research has waxed and waned since the 1960s, although it should be borne in mind that focus groups were a market research tool well before this.

During the 1960s and 1970s the tendency was to formalise qualitative market research which essentially comprised of focus groups and depth interviews. This could be interpreted as nervousness on the part of the client organisation. Could the client really be sure the focus groups, generally consisting of eight participants, had got it right? Would it not be better to take a more straightforward approach and establish preferences across a wider, more statistically reliable audience albeit without gaining any real insight as to why preferences were held? On occasions, the market research world split into those advocating qualitative research and those advocating quantitative research, with neither side seeing, or admitting, that both approaches had a role and that one could complement the other.

During the 1970s and 1980s focus groups, and *qualitative* research *per se*, became more accepted and respectable. This was partly because the process divorced itself from *motivational* research. The term motivational research had been frequently used in the past and it had uncomfortable associations, as far as commercial clients were concerned, due to the emphasis placed on the psychoanalytical interpretation of findings. Manufacturers and suppliers wanted to understand consumers' preferences, behaviour, needs and attitudes, but were not concerned if they were, for example, based on childhood experiences! They could do nothing about the childhood experiences, but they could respond to consumers' current behaviour. Additionally, qualitative research referred to a specific research process, i.e. focus groups or depth interviews followed by cognitive analysis based on the information collected via the focus group or the depth interview. Thus an organisation, carrying out market research focus groups, acquired knowledge on a specific topic through the thoughts, experiences and senses of consumers, in a flexible, but controlled environment. Moreover, the focus group moderator, although handling the groups objectively and going down the paths that were important to the client organisation, was able to present the research findings at two levels. Firstly, to present the empirical findings, which were the detailed comments and reactions of the participants. Secondly, to interpret and give additional insight into participants' behaviour and reactions. This is the approach generally adopted today.

As to the current position of focus groups in market research: in terms of industry turnover focus groups form a significant part being 20-25 per cent of all market research carried out in the UK, and the UK is at the leading edge of the growth in qualitative research.

Focus groups have expanded in the UK, especially in the last 10 to 15 years, because organisations outside the more traditional areas of new product development, advertising and marketing have seen their value, e.g. the press, broadcasting and politics.

3.3 THE REAL VALUE OF FOCUS GROUPS

In the context of market research, the real value of focus groups derives from the flexibility of the approach, being able to cover the same ground from several different angles, employing a holistic approach and active consumer/participant input.

3.3.1 Flexibility

The topics to be covered in a focus group do not have to be arbitrarily curtailed. Naturally, there is the 'main topic', i.e. the question or the problem around which the entire research programme is built, but within this context there are, or should be, no forbidden areas. This means that those organising the research, namely the client and the market research agency, are not forced to determine the criteria that may be affecting a market. Nor do they have to prioritise such criteria. This is invaluable, for however hard a client and/or the market research agency try, they are not 'normal consumers'. They are not truly objective and pertinent areas can all too easily be overlooked. The client and the market research agency do not represent normal consumers' since:

- the client is too close to the product, or the service under investigation. Moreover, the success of the company, or even individuals within the company could rest on the findings of the research, for 'politics' are often not far removed from market research. Thus, on the client side there is 'too much knowledge', too much awareness of the relevance of specific courses of action;
- the market research agency is too used to the process of 'evaluating' products and services. This does NOT mean that they can pre-judge which products and services will be successful. What a market researcher does know is the wide range of methods consumers can bring into play when faced with a new concept. Thus the researcher is not fazed by any approach adopted by a focus group, however illogical. Moreover, the route taken by a focus group when assessing, for example, a new product, can well be indicative of underlying reservations, even if the participants

state they are impressed with a product/concept. The experience of the market research agency and particularly the focus group moderator will enable this to be picked up.

As to objectivity: in the confines of market research focus groups there is probably no such thing as true objectivity. Moreover, it is unrealistic to expect it to exist. This is because focus groups are made up of groups of individuals who, unwittingly, express opinions that are based on past experience, current circumstances and personal aspirations. Nor can objectivity be expected from the client. Taking two extreme examples:

- a client developing a new product/service is naturally hoping it will be well received, that it will take the client organisation forward and to differentiate it from the competition;
- a client monitoring the effect of a newly launched competitive product/service will, not unexpectedly, be hoping that its comparable product/service is not going to be adversely affected.

How does the market research agency fare with respect to objectivity? It is the role of the market research agency to carry out the focus groups in a straightforward, unbiased manner, extrapolating as much information as possible and separating out the *facts* as presented by the participants, and the *underlying meaning* as expressed by the agency's interpretation of the facts. Thus the market research agency acts as a buffer or mediator between the consumer, the end-user and the supplier or developer of the product/service.

To be successful in this role, the market research agency has to communicate clearly and accurately the findings of focus groups, but in such a way that if reactions are highly negative the morale and confidence of the client organisation are not damaged. One way of achieving this is to suggest development routes to overcome the problems. This involves a proactive approach on the part of the agency whilst taking the focus groups.

If, on the other hand, reactions are extremely positive, the market research agency has to ensure that the client organisation does not run away with the feeling that all potential consumers will accept the product/service, and that it will effectively sell itself. In today's demanding world this is highly unlikely. To overcome this potential reaction, the agency has to act as devil's advocate and seek out any underlying negatives, any scenarios that could dent an otherwise highly favourable reaction. This has to be explored when the focus groups are being carried out and when reporting back to the client.

Thus the agency is not acting in a totally objective way, but more in a 'responsible' manner looking for ways in which products and services under investigation may be improved and sales maximised and sustained. This highlights the root cause for carrying out market research focus groups, which is to reduce commercial risk by:

- highlighting the most effective platform for launching and sustaining a product/service;
- indicating future product/service development so the client maintains any leading edge it has;
- helping the client differentiate itself and its products from the competition.

3.3.2 The Same Ground from Different Angles

Within market research, the prevailing view is that training in psychology is not necessarily a prerequisite to becoming a focus group moderator. There is no doubt, however, that human psychology has to be taken into account when designing focus groups, carrying them out and interpreting the results.

It is human nature to want to be seen in a favourable light, although perceptions of this can vary considerably by target audience. This aside, there is a tendency for focus group participants to avoid putting forward opinions that are potentially embarrassing either to themselves or to the group as a whole. In the main, participants want to be helpful. They wish to seem intelligent, sensible, well adjusted, and in reasonable control of their lives, without being too neurotic. This is not confined just to focus groups. It is the behaviour of most of us in any social situation. However, it can result in product/service negatives being played down and participants giving the impression they are more organised than is really the case.

Few participants actively or intentionally lie, but almost all subconsciously modify the degree to which they do or feel something. This can have considerable bearing on reactions to a product/service. To overcome this it is vital that the same topic is covered over a period of time from several different angles. The focus group environment is ideal for this. However, it has to be carried out with care and flair on the part of the moderator, for the participants must not get the impression that they are being 'checked on'. When analysing a focus group, the reactions/responses to a specific question/topic can be tabulated as the group progresses. By taking into account the context of each coverage of the question/topic, a more realistic and detailed picture is built up of actual behaviour and the participants' depth of liking of a product/service.

3.3.3 Holistic Approach

In market research, clients focus on particular products/services, but the consumer does not! The consumer develops a lifestyle in which many products/services play a part. The part played by each product/service can vary even on a day-to-day basis. The factors affecting this include life stage, mood, fashion, personality, peer pressure, income (especially disposable income), availability and convenience. These factors can be summed up as the needs, beliefs, behaviour and attitudes of an individual, both overall and within specific contexts. This highlights the dynamism of the consumer's world. Focus groups are invaluable in not allowing suppliers and market research agencies to forget just how dynamic the consumer is. They are also invaluable in giving insight into the way a market is moving overall, and how the role of products/services within it and bordering it are being modified or need to be modified. This is extremely important because the consumer has only so much disposable income and needs have to be prioritised. This necessitates establishing detailed background information about a product's target market.

In an ideal world, clients should, on a regular basis, carry out market research focus groups exploring the lifestyles of their primary and secondary target markets. When this has been done, possibilities for product development have been identified and real market opportunities have been realised. Many of the new products developed in the 1960s, and which are now household names, originally arose from such research, for example, fish fingers and fabric softeners. More recently, in the 1990s, Morphy Richards (a British manufacturer of mainstream consumer products) was looking to develop an electric deep fat fryer. Focus groups amongst users and lapsed users of electric deep fat fryers highlighted that a major problem with currently available products was cleaning. Ideally, the participants wanted a deep fat fryer where the bowl/tank used for frying could be removed and conveniently washed. Technically, this was difficult to achieve, but Morphy Richards persevered and were the first to introduce a deep fat fryer with a removable tank. The relevance of the move and the extent to which it met with consumer needs was strongly reflected in the high sales achieved.

Figure 3.1 Deep fat fryer with removable tank, developed in response to findings from focus group research

Establishing relevant background information is not confined to adult focus groups. Staedtler (UK) Ltd (a manufacturer of office and stationery products) recently carried out a programme of focus groups covering children and adults. The aim was to establish awareness, knowledge, usership and attitudes to writing instruments which included pens, pencils, coloured pencils as well as erasers and pencil cases. Not surprisingly the needs of adults and children differed, but it was interesting to note that when it came to coloured inks and achieving different *looks*, adult women were not that different from children in the extent to which they enjoyed exploring all possibilities.

Nowadays, time and money is at a premium. Market research projects, including focus groups, are far more focused and there is a reluctance to collect background information. However, the experienced market researcher can overcome this to some extent since they normally cover a broad spectrum of markets. As a result, they see the same target market in a range of guises, for example: doing the main supermarket shop and deciding what fresh and pre-prepared food products to buy; planning their leisure, be it a holiday or a trip to the cinema; buying children's clothes or something for themselves for a special occasion; going through the trauma of buying a home, or breaking the self-imposed diet and having a large bar of chocolate! Over time the researcher learns how the pecking order of needs and attitudes change, often resulting in seemingly conflicting behaviour. This can be fed into focus groups so that a more comprehensive picture is obtained regarding the subject under investigation. Thus the researcher has to adopt a holistic approach when carrying out focus groups, without the main purpose of the focus groups being lost or minimised. This means there must be sufficient time within a focus group to cover:

- the background of the product/service under investigation, for example, usage behaviour, awareness and knowledge of the main suppliers;
- image of the main suppliers/brands, product awareness and understanding;
- level of interest in the product area.

Only then can any new product development be assessed realistically.

3.3.4 Active Participation

Adult consumers are no different from the researchers and the clients organising the focus groups in which they are taking part. Thus they are just as capable, perhaps even more so, of evaluating products and trying to determine ways in which they can be improved. This talent must be unleashed, where appropriate, in focus groups. Projective techniques are useful tools (see methods such as *product personality profiling* and *association* – Section 12.4), but the value of interaction between participants must not be overlooked. If this is handled well, a focus group moderator will appear to almost let the participants take the focus group! Such a focus group moderator will just gently nudge the conversation down the most pertinent route. The participants will ask and answer the specific questions themselves. This increases the objectivity of the focus group in as much as the moderator is on the sideline and facilitates the natural flow of product/service assessment.

3.4 MAXIMISING ON FOCUS GROUPS

In the context of market research, it is most important that close attention is paid to the research brief, the stimuli for use in the focus groups, participant recruitment, the moderating guide, data analysis and reporting.

3.4.1 The Research Brief

The most useful research briefs explain the background to the project, the current state of play and the way in which the client/supplier wants to go forward. They also explain the pros and cons of the product/service under investigation from the client/supplier point of view and the potential marketing stance, i.e. brand, price and promotional activity. This enables the researcher to take on board the client's thinking as well as reservations underlying the product/service. This is important because the basic aim of market research is to reduce or control the risks involved when marketing products and/or services. Only by having a picture of the current state of play can the risks be anticipated. Additionally, the research brief forms a viable structure when reporting back to the client, and in its preparation the client has to be as objective as possible about the commercial moves under consideration.

3.4.2 Research Stimuli

These can range from 'concept boards' consisting of a simple sentence that blandly states the basic premise of a product/service through to working models of a proposed new product. When choosing appropriate stimuli, there are two main considerations. Firstly, the basic aim of the stimuli, e.g. to communicate a product idea, to stir up controversy and hence a more in-depth approach to a problem, or to provide a way of developing a picture of a product's target market. And secondly, the extent to which the stimuli should be 'finished.'

If the aim is to communicate a basic idea, stir up controversy, or help develop a picture of the target market, concept boards with written statements and/or associated illustrations are appropriate. Within this context a relaxed almost 'home-made' look is appropriate as it can convey the feeling that the idea being communicated can be changed; that it is not written in tablets of stone. A home-made look can actively encourage participants to suggest alterations and revisions and so opens the way for the moderator to explore new routes. However, if a concept has, in effect, been finalised and the aim of the focus group is to check on concept validity and communication, then a 'finished' approach must be adopted so that a more final and complete entity is conveyed. Whichever is used, the wording must be clear and easy to read, and the illustrations easy to interpret. The extent to which participants challenge such stimuli, both spontaneously and after prompting, is useful in gauging the credibility of the concept. All too often, insufficient attention is paid to concept boards. Far more can be gained from them if they are carefully considered when the research brief is discussed between the client and the research agency.

More finished stimuli come into their own if new products are being investigated. The ultimate form is a model or representation of the proposed new product. The value of models/representations over illustrations, photographs or technical drawings cannot be too strongly stressed. They are vital if a product has a tactile or handling aspect. It is extremely difficult for most people to accurately interpret drawings and illustrations. Furthermore, their use results in a loss of control, because it is impossible to know exactly what each participant is envisaging. Thus any conflicting views expressed could be due to widely different mental images and not to the product *per se*. Product areas where the development of models has considerably increased the accuracy of research are small electrical appliances, cleaning products, storage boxes and china tableware.

Wooden and, to a lesser extent, foam models have been used when investigating electrical appliances such as kettles, toasters and irons. When carrying out work in this area for Morphy Richards, initially drawings and photographs were used. However, the handling of such products was so important that models were developed. Admittedly, wooden models tend to be heavier than the final product, but they provide a meaningful indication of consumer reactions as regards appearance, handling, ease of use, safety, and even the position of the product within the market place when set alongside the potential competition.

When working with the Addis Group Limited (manufacturers of kitchen and domestic products) during the development of their Ultima mop, working models of the proposed product were used. As a result, the consumers' initial understanding of the product and the features that had to be communicated were quickly learnt and the development process was speeded up. The same applied to the development of Addis Lunchtime, a storage box targeted at those who prepare food to be eaten away from home. By having models to hand at all stages of the market research, the opening and general handling of the box could be observed and participants' reactions more readily appreciated. Additionally, the participants could more accurately state if the proposed product fitted with the type of food they ate away from home.

As to china tableware: when first carrying out market research for Royal Doulton (manufacturers of tableware), drawings and photographs of proposed product designs were used. However, tableware is highly tactile and this element was not present via drawings and photographs. Additionally, a pattern, when applied to tableware, takes on a life and a character of its own. Again this was missing when illustrations and photographs were used. As a result, Royal Doulton gradually developed a process of fixing the proposed pattern/design onto actual china. Although the glaze cannot be reproduced, the resulting test products have, on occasion, almost fooled participants and so the research environment has become more realistic. This has contributed significantly to the value and reliability of the research.

Models are expensive to produce. This can be a problem if there are several versions of a product. However, if there is a model of the basic product, product variations, especially if they are limited to colour and/or graphics, can be successfully covered by good quality photographs.

3.4.3 Participant Recruitment

The structure of focus groups is vital. The research brief should indicate the proposed target market and this must be represented. However, a check should be made that the target market is comprehensive. This often necessitates including participants that are to the edge of the target market, or even some distance away. This aspect is becoming more and more important as consumers become more flexible, live longer and refuse to give in to old age. In the 1960s consumers slotted more readily into set scenarios. Certainly those aged over 60 years tended to stick with a particular brand or product. The same is not true nowadays.

When structuring market research focus groups the following criteria have to be taken into account: age, sex, social class, family composition, lifestyle, current usership and location. Other factors that need to be considered are personality and the extent to which a participant is innovative and a leader in terms of taking up new ideas and products. However, it is vital that the focus groups are not over-defined since they will not reflect the eventual target market at which the product, the advertising and promotional strategies are aimed.

3.4.4 Moderating Guide

Having studied and discussed the research brief and finalised the structure of the focus groups, the research agency draws up a moderating guide. At Davis Ives two guides are developed. The first is very detailed and effectively explores all the topics to be covered. To some extent it is almost a written brainstorming session except that possible answers are not included. From the research agency's point of view, it is an extremely valuable exercise because it forces the researcher to absorb the problems and challenges facing the client and to anticipate the likely range of responses from the focus group participants. Thus the researcher is well equipped to explore all relevant avenues. This detailed moderating guide is normally included in the research proposal. It enables the client to check that key questions and areas have been absorbed by the agency. However, such a document is difficult to work from in a focus group situation. Moreover, a focus group rarely, if ever, follows the logical route of a moderating guide. Therefore, at Davis Ives, the moderator studies and almost learns the detailed guide, but works from an abbreviated format when conducting the focus groups.

3.4.5 Focus Group Analysis

Agencies, as well as individual researchers, have their own way of analysing focus groups. At Davis Ives all focus groups are tape-recorded and the moderator also takes notes during the course of a focus group. This is mainly because it is not always possible to identify individuals on the tape, to recall what they were looking at or record their body language. The moderator notes such aspects, for example, a participant giving a favourable view, but looking uncomfortable or even reluctant, or a participant seeming to change what they were about to say due to a comment interjected by another. Additionally, the moderator will note facial reactions as a product or concept board is introduced as well as a participant making a firm statement that he or she later totally contradicts.

Analysis of focus groups consists of listening and taking notes from the tapes, linking the notes made at the group with the audio tape-recording of the group and charting reactions to stimuli and products across the groups. This is a time-consuming process. In today's world many clients think that results are to hand as soon as the last focus group has been completed! It is feasible to do the basic work and give verbal feedback on the main points three to four working days after the last focus group, but to prepare a detailed and actionable written report takes two to three weeks. It is time well spent as the more careful and detailed the analysis, the more the client will get out of the research. An example of this concerns a new product just launched by Royal Doulton, namely, the tableware range Shore.

When Davis Ives first researched Shore via focus groups in both the UK and the US, reactions were not encouraging. As the product name suggests, the designers had taken the coastline as the underlying theme and had used motifs such as pebbles and seaweed on tableware. The method of execution was too realistic for the consumer, especially in the US. However, despite being critical of the design, participants in both countries were able to detail likely buyers, although taking great care to disassociate themselves from the product! By charting reactions to the product across all the groups, regardless of country, it gradually emerged that the underlying concept had merit, but it had to be carried out with great care. It was decided to re-work the concept taking account of the findings of the initial research. This meant almost starting at the beginning again. It was extremely tempting to move onto a new theme and not risk further rejection, but this would have involved extending the lead-time before a product was ready for launching. The revised version of Shore was put into focus groups in both the UK and the US and the reaction was favourable. The initial target market was again linked with the product, but this time the participants did not disassociate themselves. Most important, the final product was extremely well received by the trade. It would have been all too easy to have discarded the project, and a potentially successful product would have been missed if a detailed analysis had not been made of an extremely negative reaction.

3.4.6 Reporting

There are usually two reporting occasions: a verbal feedback or debrief, followed by a detailed written report. The verbal debrief is invaluable to both the client and the researcher. From the client's point of view they get a lead as to the next stage of development. Psychologically many clients feel they are being 'held up' while focus groups are in progress. This feeling is to some extent exaggerated because the client is distanced from the research process. The verbal debrief draws client and researcher together. At the start of the meeting the researcher dominates as the main findings are revealed. But very quickly the client asks probing questions and both sides learn more about what occurred in the focus groups, and the real potential of the product/service under investigation as well as possible marketing and promotional strategies. The researcher, by noting the reaction of the client, is more able to structure the detailed written report so it fits with the client's way of working.

3.5 FOCUS GROUPS – WHERE ARE THEY RELEVANT?

In market research focus groups are primarily used to:

- collect information when nothing is known about a market, a product or a problem that has been encountered;
- establish background data, e.g. the needs, beliefs, behaviour and attitudes of consumers within a market;
- investigate consumer behaviour and attitudes to pinpoint areas that need quantification;
- evaluate a wide range of products and/or marketing opportunities so that the most viable are taken forward for further evaluation;
- determine the most appropriate format for quantitative research, e.g. words that should be used in a questionnaire or how certain tasks should be described;
- explore new concepts relating to products and services – including new product development as well as the re-launching of existing products/services;
- explore marketing concepts, including advertising, packaging and promotions.

With new product development, market research commences when a product is thought to have commercial viability. Designers may well have carried out focus groups as they take a product concept forward. They are concerned with the product: how it works, ease of use and the end-users' reactions to its appearance. Market research is also concerned with this, but also considers outside factors such as brand, price and promotion. In other words, the full marketing mix has to be explored because, at the end of the day, sales have to be achieved and sales are not, unfortunately, directly related to the value or relevance of a product in absolute terms.

At this point, the proposed supplier's marketing team becomes involved as well as R&D (Research and Development) and, on occasion, sales. To designers who have spent a lot of time developing a concept into an actual product, it must seem as if their territory is being invaded; as if 'other people' are taking over. To some extent this is true, but it can be avoided if the designers are prepared to be involved and take on board the elements that come into play in this 'next stage'. An example of this working most effectively involves Addis Group Limited and their development of the Addis Ultima mop and the Addis Lunchtime (sandwich/meal box). The lead designer took an active part in the market research focus groups. He attended all the focus groups, acting as the moderator's assistant and helping to display and hand round the test products, all of which were working models. He found it invaluable watching group after group learning about the products and trying to force them into their current terms of reference. Davis Ives found it extremely useful having someone on hand who could give assistance, especially in terms of feasibility when participants wanted to look into product modifications.

Brand is one of the most important additional factors that now come into play. Brand links directly with supplier. Brands have images and identities that develop and change over time. It is most important that a new product is correctly branded. If a quality product is linked with a down-market brand or vice versa, relevant target markets will be confused and/or disappointed. There will also be a lack of credibility that will result in lost sales. Focus groups are an ideal environment in which to explore brand.

A new product can aid a brand. Many companies are keen to expand their brand image. In fact, brand 'elasticity' has concerned many companies over the last few years. For example, as Morphy Richards extended its range of electrical appliances, it was necessary to study the extent to which the brand fitted with the new products. When this was not the case, the company had to modify its marketing and overcome any shortcomings noted by consumers. An example of this concerns vacuum cleaners. When first considering this market, focus groups indicated that Morphy Richards, as a brand, lacked the necessary product expertise. To overcome this, the company highlighted product advantages that showed an understanding of consumer needs as well as attention to detail, e.g. the high filtration factor available on some models. Recent focus groups carried out for Morphy Richards show that this move has been successful. Now the brand has a foothold in the market. This highlights the dynamic nature of a market, the ability of brands, products and marketing to bring about real changes.

Price is another complex issue that has to be investigated. Focus groups are normally the first stage of investigation because price naturally arises during the course of new product development. The price structure within a market can present problems, mainly because the new product comes up against the competition. The standing of the new product has to be gauged in absolute and comparative terms. Focus groups provide the ideal environment to do this. Participants can examine the new product in isolation and determine what they feel to be a realistic price that they can compare with their recall of the competition. Then the competition can be physically introduced into a focus group, care being taken that a broad price range is covered. Participants assess the individual products and effectively 'map' the market by grouping and clustering the products displayed using criteria that they deem pertinent (see *conceptual mapping* in Chapter 12). The new product is positioned within this map. By getting participants to expand on the pros and cons of the different market categories, for example, price and the perception of good value, insight is gained as to how the new product is likely to fare in the market place in the context of price.

Focus groups can also play an active part in the development of promotional strategies. A useful way of achieving this is to ask focus group participants to work together to draw up the strategy they would adopt if they were the manufacturer, and then present this to the moderator. Alternatively, a focus group can be divided into two teams, both of which devise a promotional strategy, which they then present to each other. As the two teams challenge and question each other, even more is learnt about the proposed product and relevant marketing activity.

3.6. THE FUTURE

For the design and the eventual launching of a new product, focus groups are a vital tool for all concerned with the development of the product (the designers) and those more involved with the marketing aspect (the manufacturer and the market research agency). In order to get the maximum out of focus groups, everyone involved needs to adopt an integrated approach, with the role of the individual players being well defined and clearly understood. However, when the market research focus groups are carried out, it is imperative that a flexible approach is adopted. Focus groups are a tool and they have to be constantly adapted to fit with the ever-changing consumer. Providing they are flexible and explore relevant areas in depth, they will continue to play an important part in bringing new products to the market place for the foreseeable future.

3.7 RELATED READINGS

Gordon, W., 1999, *Good Thinking: A Guide to Qualitative Research*, (Henley-on Thames: Admap Publications).

Marks, L. (ed.), 2000, *Qualitative Research in Context*, (Henley-on-Thames: Admap Publications).

Sampson, P. (ed.), 1998, *Qualitative Research Through a Looking Glass*, European Society of Opinion and Marketing Research, (Amsterdam: ESOMAR).

CHAPTER 4

Focus Groups in New Product Development: Designers' Perspectives

Anne Bruseberg and Deana McDonagh

4.1 INTRODUCTION

For designers, knowledge of user needs and aspirations is vital for effective design solutions. Incorporation of user needs into designing may lead not only to commercial success, but also, it is hoped, to social gain for the users of the new products (e.g. through more satisfactory products). User-centred design approaches aim to expand the designers' knowledge, understanding and empathy of users.

One of the primary barriers to be overcome by designers is the acceptance that they need to be aware of their own limitations. Their background, training, age and possibly gender may make them distinctive from their target users. This is not to say that a 20-year-old female designer cannot respond to the needs of a 75-year-old male consumer, but to highlight that the boundaries of one's understanding can provide the impetus to seek further understanding through design research. Designers need to widen their *empathic horizon* through contact with users and exploration of user needs, as visualised in Figure 4.1. By seeking to comprehend aspects such as the lifestyles, cultural values and experiences of the target user group, designers are better equipped to empathise with the needs and aspirations of prospective users.

Design research methods and techniques place designers' own understanding in context, and offer a basis for user-centred designing. Moreover, products do not exist in isolation. They satisfy a number of users' needs beyond the functional. Thus, practising designers benefit from the availability of applying methods and techniques to elicit these often intangible needs during designing. Designers may benefit particularly from the terminology of the users.

Design research, within the discipline of Industrial and Product Design, is an emerging culture that is increasingly integrating methods and techniques from other disciplines (e.g. human factors/ergonomics, social sciences and market research). Conventional design training has, until relatively recently, not incorporated design research methods and techniques within mainstream Industrial and Product Design training programmes. Through a number of case studies (e.g. McDonagh-Philp, 2000), it has been recognised that designers would benefit from the inclusion of design-research methods at undergraduate level. In a study investigating the skill requirements for Industrial Designers, recently graduated practising designers considered themselves as 'poor researchers and would now view researching skills as an important aspect of design education' in relation to their creative problem solving skills (Garner and Duckworth, 1999, p. 94).

Designers need to immerse themselves into the design task specified by the design brief by collating a range of different types of information. Often, designers do not have access to the original primary data collected by market researchers. Through this division of activities, vital details may become lost because results tend to be summarised.

Designers possess a relatively unique range of skills, sometimes referred to as their 'toolkit' (e.g. lateral thinking, ability to innovate and visualisation skills). Focus group techniques offer designers an addition to their existing toolkit, by acting as either moderator or observer. Focus group discussions can provide designers with an insight into users' needs, aspirations and emotional bonds with products.

This chapter explores practising and training designers' perceptions of focus groups, with particular emphasis on the benefits and drawbacks, and concluding with recommendations for overcoming these barriers. It draws on three case studies.

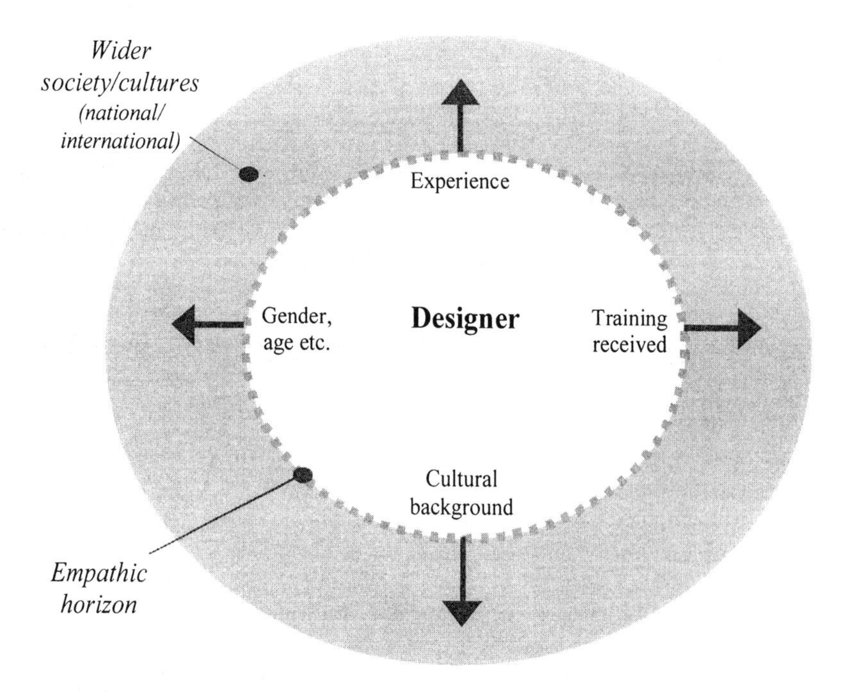

Figure 4.1 The empathic horizon of designers (McDonagh-Philp and Denton, 1999)

4.2 DESIGN RESEARCH AS PART OF THE DESIGNING PROCESS

Each design project tends to be unique in nature, due to variables such as the scope of the designing task, personnel available, client, budgets and time constraints. Bearing this in mind, each project will follow a unique process. Designing is not prescriptive. Figure 4.2 illustrates a basic range of techniques utilised during designing.

On receiving a design brief, designers (the design team) will *explore* an identified problem, product opportunity, user, and use environments, within a context. Through the use of brainstorming and mind mapping, designers are encouraged to suspend reality and think laterally. Such activity helps to generate triggers, novel uses of materials, and hopefully unexplored solutions and possibilities. This stage requires a non-judgmental environment to encourage creativity. A mood board may be created as a direct emotional response to the brief, and often after the initial brainstorming phase. It is a collection of abstract images that offers designers cues to colour, texture and form. Designers can use these images to immerse themselves into the emotions that the future product is hoped to generate in purchasers and users.

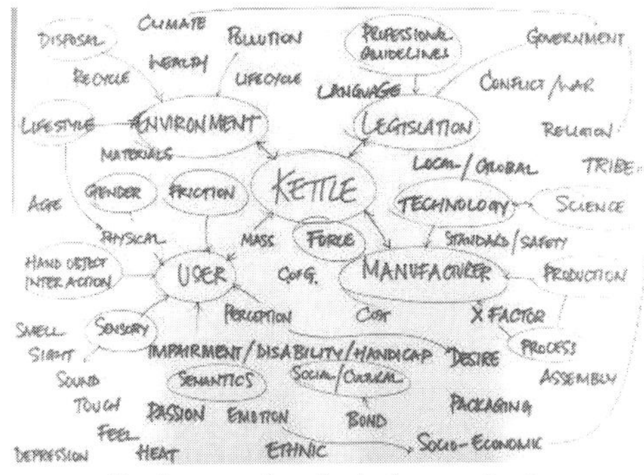

Brainstorming (mind mapping)

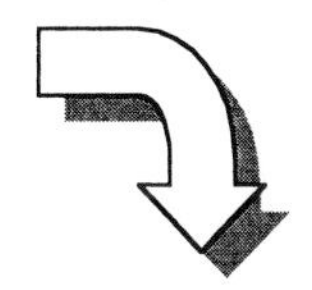

Mood board

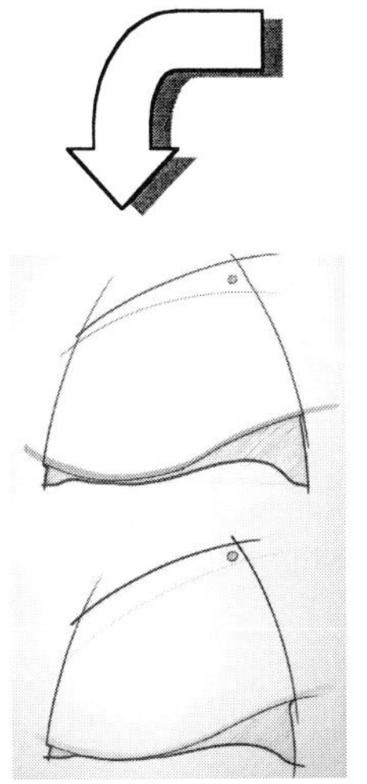

Initial concept generation drawings

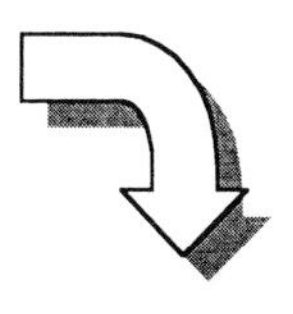

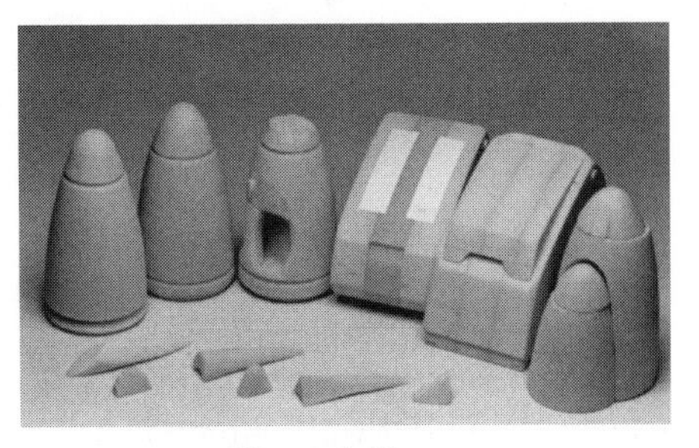

3D modelling

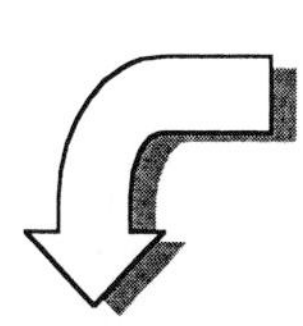

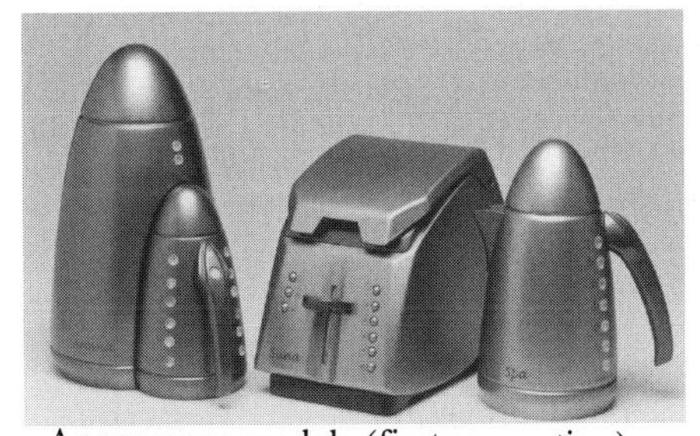

Appearance models (first generation)

Figure 4.2 The designers' basic toolkit

The next stage, *concept generation*, involves exploring form and visualising concepts two-dimensionally. Once a range of concepts has been selected for further *concept development and refinement*, three-dimensional models are generated. This enables a more thorough evaluation of the concepts by users. Appearance models are produced throughout the designing process to create an almost realistic concept proposal for evaluation.

Figure 4.3 visualises how designing activities (shown on the left) may be combined with user research activities during the concept designing stages. At the beginning of a project, it is beneficial to concentrate on eliciting user perceptions, experiences and wishes. More attention may be paid to aspects such as user lifestyles, dreams, user tasks, user environments, situations of product use, user experiences, the evaluation of existing products, and/or the use of materials. The activities should aim to increase the understanding of user needs beyond the functional (e.g. product styling). It is useful to 'step back' and consider the entire product context (e.g. user tasks and environments of product use). This may reveal needs that have not yet been identified. Users may be invited to brainstorm about ideal product solutions, either verbally or through creative activities, such as drawing or modelling exercises. Since focus groups retrieve qualitative data, they are particularly suited to these functions.

After the initial concepts have been generated, user responses may be sought to assist the selection of concepts for development and further refinement. Concept evaluation by users may involve the assessment of ideas, drawings, models and prototypes. It is helpful to select a diverse range of concepts (ranging from incremental changes to existing forms through to more abstract, blue-sky proposals) to be presented to users and to provoke responses. Early concept evaluation sessions may still focus on further idea generation. They aim at retrieving feedback from users on the designers' perception and understanding of user needs. The concepts enable further exploration of users' wishes and can be used to inspire users' brainstorming.

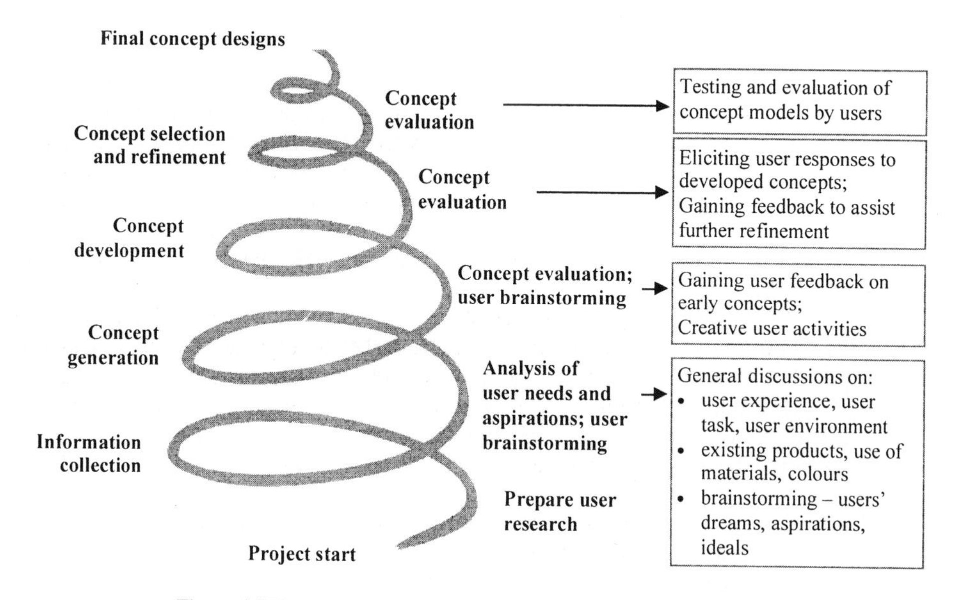

Figure 4.3 The interaction between industrial design and design research

In practice, manufacturing companies tend to involve users at the concept evaluation phase at the end of the concept designing stage. Focus group research is able to provide a substantial knowledge base about user needs and aspirations before designing activities commence. This approach helps to avoid design fixations (Jansson and Smith, 1991). If designers begin concept generation prior to being immersed in the user experience, they tend to invest heavily in the early concept work. Later research findings may indicate a need for a change in direction. Due to this early investment, designers may become resistant to alternative avenues of exploration. Once concept generation has began, the research and designing process should continually inform each other in an iterative manner (see Figure 4.3). As there is only little scope for comparison between sessions, the contents of the sessions may be varied according to knowledge gained previously.

The level of involvement of designers during the research activities may vary depending on the time, resources, personnel and the budget available. The authors envisage three different scenarios for the level of responsibility that designers may have in conducting the user research:

Scenario A: designers carry out all the associated activities;

Scenario B: designers delegate administrative and organisational tasks;

Scenario C: designers are only partly, but actively, involved (e.g. preparation of session contents, co-moderating) and collaborate closely with design researchers.

4.3 DESIGNERS' PERSPECTIVES

4.3.1 Three Case Studies

Two studies with five design practitioners and seven undergraduate design students have been conducted to explore designers' perspectives of focus group techniques. The practising designers' ages ranged from 22-52 years and included two freelance designers, one working for a design consultancy and two working as in-house designers for a manufacturer of consumer products (see also Bruseberg and McDonagh-Philp, 2000). During one-to-one interviews, the designers provided feedback and comments on the appropriateness of focus group techniques in relation to their own needs and designing processes.

The students' ages ranged from 20-22 years, and were a mix of second year students and finalists on the Industrial Design and Technology degree programme at Loughborough University. The majority of them had recently returned from industrial placements. The students' feedback was retrieved in the setting of a focus group after being given a detailed presentation with drawbacks, benefits of focus groups, practical requirements, and the various techniques that may be integrated (refer to Chapter 2, Table 2.1).

The third case study involved a practice-based design-research project. A range of user research techniques based on focus groups was applied during a design project for small domestic kitchen appliances. The authors worked as the design researchers in close collaboration with a consultant designer, who provided valuable feedback regarding the techniques used, the information provided, and the design team collaboration during a one-to-one interview.

The findings presented here do not aim to show evidence for general designers' thinking, but show the perceptions of the designers interviewed in this study, thus representing a *sample* point of view. Moreover, they are summarised here to give a brief overview, rather than explore the issues in depth. Further detail can be found in Bruseberg and McDonagh-Philp (2002).

4.3.2 Practitioners

Previous experience

Although the designers had a basic understanding of focus group activities, none had been involved in focus group research. Two designers from the manufacturing company had once observed a focus group session, which had been organised by a professional moderator. For most designers, focus group data were a rare source of information. If there had been contact with focus groups, then it was usually in relation to market research.

Focus groups and the designing process

Designers stressed that designing is a flexible process that draws on a large amount of information and the designers' experiences. It is an intuitive, unpredictable and continuous process. Designers find it difficult to conceptualise their own designing process and can rarely explain what really drives ideas to occur. Designing draws from a large variety of resources.

The methods used vary considerably because of the differences in design areas as well as project circumstances, but also because of individual differences between designers. Every designer adapts a set of techniques based on training and experience to suit personal preferences as well as the circumstances of the designing task.

Designers rarely apply formal approaches or systematic methods as suggested by the design literature – partly due to time constraints, but also because the methods need to be flexible. Formal methods are considered to be artificial and have not necessarily proven to be beneficial.

Designers' reservations towards focus group activities

Freelance designers anticipated particular difficulties with finding the time to carry out user research themselves, particularly as clients are not prepared to pay for such activities. Designers are often not in control of which or how user research is to be carried out. Some of the designers perceived that there is a gradual shift in the design and manufacturing culture towards the acceptance of more user research.

Focus groups may detract designers too much from actual designing activities. A common perception by both designers and clients is that the function of the designer is to produce quick sketches, rather than examining the design problem in depth.

Consulting users was perceived as a 'weakness', as designers are expected to 'shift culture' and create innovative solutions. There was resentment towards the prospect of changing the role of the designer when incorporating a number of new roles, such as moderating or communicating with users directly.

Some designers displayed a resistance to interacting directly with users. Designers sometimes expressed concerns about conflicts of opinion with users, or lack of 'similar beliefs', particularly when being present in a concept evaluation session. There was a concern that users may not know their own needs or aspirations, or may not be able to articulate them. Designers doubted that users would be able to think creatively, both due to lack of training, but also because they cannot immerse themselves into the design task as deeply as the designers are able to.

The lack of understanding of focus group research and the potential benefits was evident. There seemed to be a perception from experience that focus groups are mainly used to evaluate concepts. There was concern about the lack of skills available to prepare and conduct focus groups, including being able to ask appropriate questions.

Designers' views on benefits from being involved in user research

The consideration of the users' tasks and the user characteristics is instrumental to designers. Whilst some designers displayed some resistance to changing their working patterns, others were more receptive provided they were given a supportive environment. Some designers stressed the importance of carrying out user research prior to concept generation; others saw more benefits in evaluating early concepts. Designers emphasised that the objectives need to be defined clearly. The outcomes should be well tailored to the information needs of designers regarding users' aspirations and needs.

User research and designing is not always sufficiently integrated. The information received from market research departments is often perceived to be unsuitable because it does not aid the creative process or may be perceived as restrictive. The presentation of the information in a summarised form and through written documentation is often resisted. Several designers expressed that they would value input into the planning of user research.

Information about user experiences, user tasks, problems, user values, and user perceptions was considered as particularly important. Cultural issues, including insights into people's lifestyles or 'what a product stands for', were considered to be of particular value for designing. Brainstorming activities with users about the 'the history of objects', as well as product trends, were regarded as beneficial exercises. Beyond that, discussing initial design ideas with users was seen as useful. Designers advocated the incorporation of techniques that provide visual information, such as observational techniques. A friendly atmosphere to support such collaboration was considered crucial.

4.3.3 Undergraduate Feedback

Students were open towards user-centred design approaches. They had mainly used questionnaire and/or interview techniques to research user needs during their degree course projects. Some students had positive experiences from involving users into their designing process during the requirements capture stage, and were receptive to spending a considerable amount of time on user research. Investigating and empathising with user needs was seen as part of the preparation for creative problem-solving. Others had encountered difficulties (e.g. time available, usefulness of information gained, lack of user co-operation) and were unsure about the benefits.

Some students had encountered a conflict between being truly innovative and drawing on users' perceptions that they regarded as restricting their creative freedom, or being 'out of date'. On the other hand, students recognised that users often have unarticulated wishes and may be particularly suitable to provide unbiased, critical feedback to design concepts. Communication with other people (e.g. non-designer friends, colleagues, users) about user needs and concept ideas was perceived as an important tool to examine the design task and stimulate the mind. Moreover, designers often find themselves in a situation where they have to defend their solutions to clients – therefore the availability of evidence from user research was seen as beneficial.

Students welcomed the semi-formal nature of focus group activities. A basic degree of formality provides a neutral environment where honest input is encouraged and sensitive issues may be tackled. At the same time, a friendly and informal atmosphere puts everybody at ease, promotes user co-operation, and supports creative activities. The group synergy (i.e. the sharing and comparing of ideas and experiences amongst group members) was recognised as valuable for retrieving useful information.

Students emphasised the need for support in formulating appropriate questions – to retrieve the required insights, promote discussion, avoid bias, and lead skilfully from one question to another. The incorporation of a variety of techniques (e.g. questionnaires, product handling and product personality profiles) besides the actual discussion was welcomed. Insights into emotional and aesthetic values were seen as important.

Students were not confident about the prospect of moderating group sessions themselves, even with training, but welcomed the opportunity to co-moderate sessions and actively take part. The time required for the analysis of qualitative data was perceived as problematic. Regarding data presentation formats, preferences were expressed towards visual information and overviews of users' values.

4.3.4 Consultant Designer Feedback on the Design Research Project

The consultant designer preferred user research involvement from the earliest stages, including preparing the research and collecting relevant material (e.g. future trends, history of products). The presence of designers during the sessions was seen as vital. Particularly during the stages of concept evaluation, active involvement of the designer was regarded as essential, such as introducing the concepts, leading the concept evaluation discussion, and responding to participants' questions. The designer benefited from previous experience in leading a group discussion – to be able to deal with situations where users are not sufficiently communicative, and to make sure all participants provide input.

Owing to the time constraints of the project, the data analysis was largely carried out by the design researcher/ergonomist. The designer valued a summarised presentation showing categories of user values, preferably in the shape of a bullet point list, but supported by a range of examples (e.g. users' quotes or selected video clips). A list containing positive and negative adjectives for selected products, taken from user product evaluations, was particularly appreciated. These provided an overview of how users perceived the styling of a number of sample products (see Chapter 2, Figure 2.18). The availability of a range of different types of data from a focus group session was perceived as useful. Besides the direct verbal feedback on early concept drawings, the designer particularly valued opportunities to observe user interaction with products, and users' expressions through drawings.

The designer appreciated being able to concentrate on studying information, preparing their input into the concept evaluation sessions, and generating/developing concepts, as all organisational activities were carried out by the researchers – such as recruiting participants, scheduling the sessions, conducting the session. The designer regarded the regular consultations with the design team as essential – to set out issues for the next focus group sessions, discuss the design concepts, review whether the concepts met user requirements, and to decide on the selection of concepts to be taken further.

4.4 CONCLUSIONS

Whilst focus group techniques have a number of limitations, several of the reservations from designers stem from preconceptions based on the lack of knowledge about the potential benefits, and from previous unfavourable experiences. Other reservations are based on practical restrictions and constraints, often due to external circumstances, such as lack of time or client attitude.

The authors believe that users need to be recognised as an indispensable design resource, as the merits of their input are often undervalued. It needs to be appreciated that the immersion into user aspirations from early on in the designing process is vital to successful product designing. Designers' comments have confirmed that considering the user is central to all design tasks. There are signs that the benefits of user-centred designing are becoming more accepted by designers, clients, and manufacturers.

Focus group techniques need to be adapted to the specific requirements of designers and creative design processes. The information provided from market research departments and/or ergonomists is often inappropriate for designers, or may be perceived as restrictive. The techniques need to take account of (a) the working environment/ conditions of designing, (b) the information requirements of designers, and (c) the needs for interdisciplinary collaboration. Design research techniques are intended to build on designers' current skills and expertise, rather than transforming designers into ergonomists or market researchers.

Designers often resist formalised approaches. Focus group research provides a basic structure to the elicitation of user needs, but also provides an adaptable method that includes a high level of informality. Conducting focus group research may thus be particularly suitable for integration into the information search that is part of designers' creative phases. Moreover, focus group techniques are useful to aid collaboration between different professionals (Bruseberg and McDonagh-Philp, 2001).

Design research is a relatively recent development within the design discipline. For practising designers this paradigm shift becomes challenging, for the student designer the design research culture is introduced earlier and more gradually. Due to the responsive nature of Industrial Design training at Loughborough University, user-centred design techniques and methods are being integrated within the undergraduate programmes. This shift in emphasis has enabled a generation of up-and-coming designers who possess the skills and expertise on graduating to implement, adopt and employ techniques, such as focus group activities, as part of their enhanced toolkit.

4.5 REFERENCES

Bruseberg, A. and McDonagh-Philp, D., 2000, User-centred design research methods: the designer's perspective. *Integrating Design Education Beyond 2000 Conference*, University of Sussex, 4-6 September, pp. 179-184, ISBN 1 86058 265 6.

Bruseberg, A. and McDonagh-Philp, D., 2001, New product development by eliciting user experience and aspirations. *International Journal of Human-Computer Studies*, **55** (4), pp. 435-452.

Bruseberg, A. and McDonagh-Philp, D., 2002, Focus groups to support the industrial/ product designer: a review based on current literature and designers' feedback. *Applied Ergonomics*, **33** (1), pp. 27-38.

Garner, S. and Duckworth, A., 1999, Identifying key competences of industrial design and technology graduates in small and medium sized enterprises. In *IDATER conference (International Conference of Design and Technology Educational*

Curriculum Development), Loughborough: Department of Design and Technology, Loughborough University. Roberts, P.H. and Norman, E.W.L. (eds), pp. 88-96.

Jansson, D.G. and Smith, S.M., 1991, Design fixation. *Design Studies*, **12** (1), pp. 3-11.

McDonagh-Philp, D., 2000, Undergraduate design research: a case study. In *Collaborative Design (Proceedings of CoDesigning 2000, UK, 11-13 September).* Scrivener, S.A.R., Ball, L.J. and Woodcock, A. (eds), (London: Springer-Verlag), pp. 493-500, ISBN 1-85233-341-3.

McDonagh-Philp, D. and Denton, H., 1999, Using focus groups to support the designer in the evaluation of existing products: a case study. *The Design Journal*, **2** (2), pp. 20-31.

CHAPTER 5

The Use of Focus Groups for User Requirements Analysis

Martin Maguire

5.1 INTRODUCTION

Focus Groups are a cost-effective way of assessing user and customer requirements for IT (information technology) systems or products. However, specifying such requirements is not a simple process. Understanding the working or domestic activities of users that may be supported by future IT systems can be a lengthy process if carried out on an individual basis and by observation. There may be many categories of user and stakeholder whose different perspectives and requirements need to be elicited and prioritised. Thus the concept of assembling groups of users and stakeholders to discuss their current activities or ideas on a particular topic can be a simple and quick way to help establish user requirements.

This chapter examines the role of focus groups as a method for specifying user requirements for IT systems. Three distinctive styles of focus group are identified. These relate to: (1) current activities and needs, (2) new design concepts, and (3) review of developing prototypes. A series of case studies is presented to illustrate the use of focus groups for different stages of the system development lifecycle. Comments on the approach taken within each case study are given and guidelines offered on running successful requirements focus groups in the future.

5.2 FOCUS GROUPS IN SYSTEM DEVELOPMENT

Focus groups are a popular means of eliciting the views of members of the public on a wide range of issues from testing people's political views to assessing the general public's reaction to consumer products. Originally they were used after the Second World War to evaluate audience response to radio programs (Stewart and Shamdasani, 1990).

A focus group can be organised in many ways but essentially it is a group of people, invited to meet and discuss issues related to a particular topic. A facilitator or moderator guides the meeting, encouraging discussion, ensuring that the meeting stays on track, and ensuring that everyone is able to express their views. Group discussions help to summarise the ideas and information held by individual members. Each participant can act to stimulate ideas in the other people present and that by a process of discussion, a collective view becomes established which is greater than the individual parts.

The main advantages of focus groups are:

- they allow people to air their views in a natural conversational way;
- a wide variety of perspectives can be sampled quickly;
- they require no special equipment and are comparatively easy to conduct;
- people normally enjoy taking part in them.

Disadvantages of focus groups are that:

- it is not always easy to control the group or the information from it;
- peer pressure may lead to inaccurate reports;
- participants may feel that they should give socially acceptable views;
- some individuals may not get the chance to air their views or may be inhibited by other group members, particularly colleagues or more senior staff;
- some people may also not always think creatively in a group setting and prefer to be interviewed or to complete a survey form in their own time.

Despite these limitations, focus groups are still a valuable data gathering technique.

Caplan (1990) was one of the early researchers to apply focus group methodology to ergonomic design. According to Caplan, marketing research deals with perceptions of abstract concepts and requires numerous sessions with different groups to compile meaningful conclusions. In contrast, ergonomic applications that deal with better defined domains can thus benefit from a small number of sessions. The spontaneity of participant interaction can bring out valuable ideas.

O'Donnell *et al.* (1991) describe the advantages of group techniques in comparison with studies based on individuals, e.g. interviews. These are: that they allow subjects to remind one another of events; they encourage subjects to reconstruct processes; they explore gaps in subjects' thinking; they overcome the 'not worth mentioning' problem and, importantly for HCI (Human-Computer Interaction), they allow new solutions to emerge. Nielsen (1997) summarises the advantages of focus groups for HCI research as that they often bring out users' spontaneous reactions and ideas, and allow observation of group dynamics and organisational issues. It is also possible to ask participants to discuss how they perform activities that span many days or weeks: something that is expensive to observe directly.

However, Nielsen (1999) also warns that focus groups' results are a poor method for evaluating the usability of a new system prototype and should not be used as the only source of usability data. In relation to websites, he states that focus groups can often be misleading. When users are asked what they want on a website, their answers may have little relation to their actual behaviour when they go online. This is because users may say what they think the moderator wants to hear and what they think is socially acceptable. Because focus groups involve several people, individuals rarely get the chance to explore the system on their own; instead, the moderator usually provides a product demonstration as the basis for discussion. Watching a demonstration is fundamentally different from actually using the product: there is never a question as to what to do next and the participant does not have to ponder the meaning of numerous screen options.

The ideal number of people who should take part in a focus group is undefined although the upper and lower limits are possibly covered by Macaulay (1996) who, for requirements analyses recommends 4-6 participants, while Nielsen (1993) suggests 6-9. Interestingly, Henderson and Hawkes (2001) report on a study of decision-makers by subjects acting as jurors. This found that seven was an ideal number of people to provide both a representative set of jurors and led to effective decision-making. Professor Simon Garrod, who led the study at Glasgow University, found that for the large groups (10-12 people), individuals tended to take turns to 'broadcast' their views to their colleagues and there was less scope for everyone to take part. Within small groups (5-7), participants could engage in dialogue with the other members and give their opinions freely as they are relevant to the discussion. Note, however, that juror meetings are self organised while focus groups have a facilitator who has the role of leading and engaging the whole group, so arguably larger numbers are manageable.

5.3 FOCUS GROUPS IN USER REQUIREMENTS ANALYSIS

Focus groups may be used within the process of analysing user requirements prior to system design (Nielsen, 1997) by allowing the analyst to rapidly obtain the views from a range of future users and stakeholders. Focus groups will give the Requirements Engineer insights into how users think and what things are important to them (Macaulay, 1996). In this situation the Requirements Engineer can act as facilitator and adopt a fairly passive role. According to Macaulay (1996), a focus group in a requirements context can support:

- the articulation of visions, design proposals or product concepts;
- user analysis of their own problems and identification of the need for change;
- the development of a *shared meaning* of the system being specified.

In practice, such discussions are rarely open-ended. Typically, a development organisation will have a particular idea or innovation that they would like to test out on groups of potential users. The views of the users can give the organisation confidence that the idea is a good one or that it needs to be changed. If a prototype or system mock-up of the idea exists, then ideally the focus group should take place after the participants have had some individual exposure to it.

Kuhn (2000) has found that focus group sessions were very helpful in requirements gathering when applied to a home automation system. He found that focus groups initiate a creative process that can influence the whole design process with ideas, suggestions and forward-looking perceptions (which often go beyond the resources and technical possibilities available). He also found that focus groups deliver data about the context of a product that cannot be gained with surveys or individual interviews. Focus groups also help convince developers and policy-makers within technical departments to take user needs and usability functions into account. Kuhn stresses the need to carefully transcribe and code the data gathered if they are to be relied upon. He found that focus groups were unsuitable for optimising sequences of operations.

A focus group may be an approach to learn about current activities or work that can feed into requirements analysis and would take too long to study directly (Ede, 1999). However, it is a mistake to simply ask workers what they do, as they will tend to provide technical descriptions of their job functions. Ede prefers to ask people to describe their last full day at work starting from when they arrive and have their first cup of coffee. This approach gets at the activities that occur *between* and *around* the systems people use and reveal many of the problems relating to the person's job. She gives the example of a focus group that she ran with system administrators who, when first asked what they did, said things like: install, upgrade, oversee, and troubleshoot. When she next asked to describe their working day, they launched into more illuminating stories about how they could not drink coffee because they would be accosted on the way to the coffee machine with requests for help. Most carried notebooks to record end-user problems. They talked about telling users many times how to do the same simple tasks, about training courses, surfing the Internet, being paged at home and reading technical books in bed.

Clearly the style of questioning is crucial to obtaining a realistic picture from a focus group of what current activities are performed and what the real problems are.

5.4 APPLICATION TO DIFFERENT REQUIREMENTS ACTIVITIES

According to the ISO 13407 standard on human-centred design (ISO, 1999) there are five essential phases that should be undertaken in order to incorporate usability requirements into the software development process. These are:

1. Plan the human-centred process.
2. Understand and specify the context of use.
3. Specify the user and organisational requirements.
4. Produce designs and prototypes.
5. Carry out user-based assessment.

Phases 2 to 5 form an iterative development cycle where users provide information that can be used to specify their context of use and current tasks and problems. This information is used to produce user and organisational requirements. From here system designs and prototypes are developed, which are then tested to see whether the contextual needs and user and organisational requirements have been met. As a result, the prototype is modified and further tests carried out in an iterative fashion. Within this human-centred approach, focus groups can play several roles.

Group focus: Current activities and needs. Here there is a discussion of the current activities that users perform and their particular needs. This supports the understanding of context of use of a future system and helps to identify possible user requirements. Participants may also comment on the existing system they use. This style of focus group is discussed in Section 5.5.

Group focus: New design concepts. Here users discuss a possible design or new technology concept and provide feedback on how useful and acceptable they find it. This style of focus group is discussed in Section 5.6.

Group focus: Prototype review. Having used a system prototype, users will meet to discuss their use of it, any problems they experienced, and features they liked. Alternatively they may provide their reactions to a system demonstrated to them, including how well it meets their preferences and requirements. This style of focus group is discussed in Section 5.7.

A focus group will tend to concentrate upon one of these three areas but it is possible that they could be combined; so, for instance, a focus group may discuss current activities and then discuss the feasibility of a new design concept in relation to them.

5.5 GROUP FOCUS: CURRENT ACTIVITIES AND NEEDS

Focus groups provide a good way to bring users and stakeholders together to discuss their current activities or tasks either in a work, leisure or other context, for example, whilst driving, shopping or sightseeing. The facilitator will typically describe a number of current activity scenarios and ask participants to describe their experiences and problems. The following section presents a case study performed relating to studying current activities relating to organising domestic finances.

5.5.1 Case Study – Financial Services to the Home

Description and purpose

The management of domestic finances is an activity that affects the large majority of people. With the increasing range of financial services and products available, the potential complexity of this task can be high. At the same time, the movement from paper-based towards electronic transactions offers the possibility for the public to manage their finances more effectively and efficiently. However, unless electronic-based financial services are offered in a way that meets consumer requirements, they will fail to be taken up successfully.

A study was carried out to interview family groups and to study how they handle and control their domestic finances. Each group was asked to describe the financial tasks they perform, where they are performed within the house, how they store and access financial records, and what means of communication they use (e.g. telephone, letter, the Internet or visit to a branch). The main aim was to describe the whole environment for managing domestic finances. It also looked at who performed different tasks within the household. The study was intended to provide useful data as a baseline to help create and deliver financial services in the future, for example, via digital TV or other household devices.

Structure of groups

Focus group interviews were carried out with 13 families, 11 being conducted within their own homes and two being conducted at the author's own offices. The interviews included the following:

- four working single parent households with one or more teenage children at home;
- five working couples with a family, with one or more teenage children at home;
- four older couples of whom at least one was retired or close to retirement.

A further discussion group session was held with three teenagers, of 16 and 17 years old, for one hour. This focused on what factors influenced them to open a particular bank or savings account, and how they used them to save and access money, for example, via bank machines.

Session format

The researcher facilitated each family group session. The interview schedule covered the following topics:

- types of financial items held by family members;
- how they receive financial information, conduct transactions, deal with problems and possible improvements;
- use of household appliances, and whether they are used for financial tasks;
- locations in the home for managing finances, storing financial paperwork, advantages and disadvantages of these locations and possible improvements;
- possible methods of accessing domestic financial services in different locations and through different appliances;
- general views on current banking services.

The researcher brought portable video equipment to each house so the session could be recorded to supplement the researcher's first-hand note taking. However for two sessions, the interviewees preferred to be audio-recorded only. Still pictures were also taken of relevant household locations to show where financial tasks were performed and where paperwork or correspondence was kept. The locations of devices used for financial tasks and the surrounding spaces were photographed, for example, telephone, television or personal computer. The intention was to show how people liked to set up their work area for performing financial tasks as this would be potentially useful information for providers of future financial services and developers of terminal equipment. Each session lasted approximately one and a half hours.

Results

The group sessions were successful in exploring the processes that families follow in managing their finances and highlighted tendencies within specific groups. For example, the discussions showed that single parents on lower salary levels liked to be able to keep a close check on their account balances and to set aside money to pay for items and bills as they arise. The working family groups tended to seek convenient means of paying bills either by direct debit so that a bill is not forgotten, or by using telephone bill-paying services offered by the bank.

Participants also discussed the possible benefits of receiving financial services electronically. Several innovative ideas for financial services were suggested that could support banking services in the future. One person suggested that they be given a device like a pager that they could swipe with their bank card to get a balance. A portable financial monitoring device could also be useful to allow financial tasks to be carried out in different locations within the house. Several people liked the idea of receiving bank statements by email, and having a portable device (like a pager) for checking an account balance. Another suggestion was for a combined telephone/bank card.

Comments on method

The study showed that organising and running focus groups with families within their own homes could work well. It provided a way of accessing a family group in a convenient manner. The discussion group format worked quite well and after a warm up phase (where the study was introduced by the facilitator) people were happy to discuss how they managed money. However, the recruitment process had to be handled sensitively and it had to be stressed that the questions were of a general nature, and did not ask for personal financial details (e.g. how much is held in a particular account).

It was important to get each family group to agree to be videotaped beforehand and to give them the option for audio recording only. Thus it was necessary to have both types of recording equipment available. The participants were also generally happy for the researcher to video and photograph locations of devices in the home such as televisions, radios, telephones and personal computers to see how they might be adapted to receive financial services.

However, some problems were experienced. Firstly, one or two families were not initially comfortable with the idea of the discussion being held in their home and for specific locations to be videotaped or photographed. It was therefore important during the recruitment process to explain the reasons for this to take place. Another issue was that not all family members could be present, and several were absent when the researcher arrived. Clearly the researcher cannot control this and has to accept that for many families it is hard to get them all together at one time for a discussion session. However, it is important to make it clear beforehand that it is desirable to include those who do not handle family finances (and who may be potential users of new services in the future) as well as those who do. Family members also have other priorities at home and it was seen as important not to interrupt them for too long, for example, when an important football match was starting on television. Flexibility and discretion were found to be the key factors for the focus group sessions to be acceptable.

Although the main aim of the focus group sessions was to discuss current financial activities, it was natural to discuss new ideas for delivering financial services. This allowed participants to imagine new situations and to come up with new ideas. Although the discussion of new ideas was not originally planned as part of the focus group agenda, this short phase at the end of the session was enjoyed by participants and made the event more meaningful to them. Such suggestions should, of course, be recorded for input into concept development activities.

5.6 GROUP FOCUS: NEW DESIGN CONCEPTS

Focus groups can be a way to generate new ideas (Osborn, 1963, and Jones, 1980). McClelland and Brigham (1990) present an example of the use of focus groups in the design of an advanced telecommunications system for office and home workers. Firstly, a user needs workshop was held where six users discussed their communications needs and problems with respect to current tools and practices. Then six experts spent two days in a design workshop outlining five different design directions and usage scenarios to address the user needs identified in the initial focus group. Finally, the focus group with the six users was reconvened to discuss these proposals for a future system

The concept of brainstorming fits well into the activity of generating new ideas activity. Poulson *et al.* (1996) describe a method for holding a brainstorming session where a well-formulated problem is presented to a group of designers. They spend five minutes sitting quietly and writing down all the ideas that come to mind, even wild ideas and thoughts as long as they are responses to the problem formulation. Each person in turn then reads out one of their ideas from their list, continuing around the table until all ideas have been presented. As each idea is presented it is discussed constructively within the group leading to new ideas being generated. Rules for brainstorming discussions are offered by Cross (1989):

- no criticism is allowed during the session;
- a large quantity of ideas is wanted;
- seemingly irrelevant ideas are quite welcome;
- try to combine and improve on the ideas of others.

Focus groups may also be run with potential users to help determine how well a new design or technology will be accepted by end-users. Since the success of consumer products or consumer services are very much dependent on the reactions of the general public, focus groups are a good way for an organisation to determine what the public or customer reaction will be. This section describes three focus group studies organised to get participants' reactions to systems based on new technology ideas.

The different topics for each study were: (1) the use of speech interaction with ATMs (Automatic Teller Machines) or bank machines, (2) the use of biometrics (human recognition systems) for identifying people and (3) development of ATMs with special functions for traders.

5.6.1 Case Study – Speech-based Bank Machine

Description and purpose

A series of focus group sessions was held to test the idea of using speech input (speech recognition) and output as a way to support verification and voice command of bank machines (Hone *et al.*, 1998). The aim was to investigate the possibility of employing speech technology within a new generation of bank machines, and in what form and location it could be successfully deployed. It is reported that there is still a high proportion of bank customers (25 per cent in the UK) who do not use bank machines, so such technologies might be a way to encourage their use (Derbyshire, 1999).

Structure of groups

A total of 16 focus groups were run, eight in the UK (Loughborough, Leicestershire) and eight in the USA (Boston, Massachusetts). One group in each country consisted of physically impaired subjects and another of visually impaired subjects. The remaining groups were balanced by age; two young, two middle-aged and two older groups in each location. All subjects had current accounts at a bank or building society, and all used ATMs.

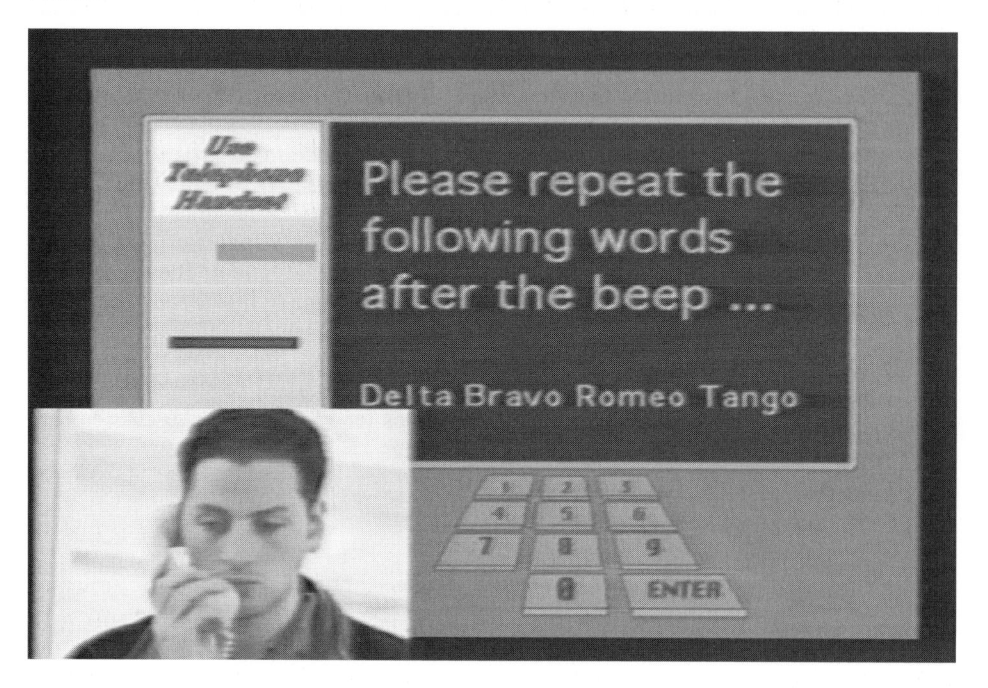

Figure 5.1 Video simulation of speaker verification process

Session format

The moderator began by introducing himself and asking subjects to introduce themselves to the group. Subjects in the visually and physically impaired groups were additionally asked to briefly describe the nature of their impairments. The moderator led the discussion by referring to one of three checklist agendas, developed for the non-impaired groups, the visually impaired groups, and the physically impaired groups. Each focus group session covered the following topics to which participants provided their reactions:

- experiences with ATMs, including problems and suggestions for improvement;
- experiences of speech input and output technologies;
- ATM security and how to improve it;
- presentation of the concept of speaker verification for an ATM covering security, reliability, comfort in using, enrolment, and the verification phrase;
- demonstration of speaker verification with two video simulations (see Figure 5.1);
- presentation of the concept of speech input and output for ATMs, covering training, speed, privacy, comfort, and preferred voice characteristics for speech output;
- practical aspects of speech-based ATMs such as the terminal type, location and type of input device.

Finally, subjects were asked to draw and describe their ideal ATM, and were encouraged to incorporate speech in their descriptions where possible. Each session lasted from one and a half to two hours.

Results

The focus group sessions revealed a general concern about the use of voice with a bank machine located in a public place, in terms of security, reliability and privacy. For instance, participants were unhappy about their bank balance being announced by voice, or the customer asking for an amount of money or speaking out loud for voice verification. Many participants felt that speech-based transactions would be slower than the existing visual-manual mode, possibly due to their previous experiences of speech technology (e.g. speech-based automatic telephone switchboards). A cross-cultural difference in attitude emerged, with American participants placing much greater emphasis on the speed of interaction than the British. There were also strong doubts as to whether the technology would be reliable enough to work, due mainly to inconsistencies or changes in the user's voice. Interestingly, a feeling of discomfort in actually speaking to a machine was a relatively minor issue. Perhaps the most positive message was that speech output could be an option, which the user could choose in order to receive instructions on using a standard ATM.

The focus groups were able to suggest a number of improvements to current ATMs. These included: increasing the level of privacy and security by placing them in enclosed/lockable booths and providing a 'panic button'; increasing the speed of interaction by reorganising the dialogue and using a touch-screen; providing human assistance at the ATM; and widening the range of ATM services and locations.

In contrast to the general views of the public, visually impaired users tended to be more positive about speech-based interaction with a bank machine. Many were frustrated about being unable to use a bank machine without help. They proposed ways in which voice could be used – with headphones that could be plugged into the bank machine. They also proposed alternative ideas that could make bank machines more usable by people who had some sight such as allowing them to select a larger font size on the screen and present it in highly contrasting colour.

Physically impaired users were predominantly negative about the use of speech with a bank machine. However, they were a very inventive group in coming up with ideas to solve problems such as the difficulty in reaching an ATM from a seated position in a wheelchair (thus demonstrating their technology awareness). One US user suggested having a pull-out keyboard that he could put on his lap to use the bank machine. Other improvements proposed were wide sliding doors on bank branches and ramps to all machines. It was also thought that large touch-sensitive keys, rather than buttons that require force, would significantly help those people with upper-limb impairments.

For a full discussion of the study results see Hone *et al.* (1998).

Comments on method

Prior to running the focus groups, a survey had been performed in the UK and USA to ask people for their reactions to the idea of a speech-based ATM. This provided a larger scale quantitative survey and helped to identify the issues to discuss in more detail within the focus groups. In the survey, responses to the idea of using speech for ATM user verification were fairly well-balanced between those who were positive and those who were negative. It is interesting that the responses of the focus groups were more strongly negative, possibly with the opportunity to reason through issues with fellow participants.

Reactions to the use of handsets to speak to a bank machine were also interesting. While shared handsets are in wide use in offices and telephone kiosks (until the growth in the use of mobile phones), it was found that when one or two people expressed concern about handset cleanliness, others became more sensitive about the issue and agreed. This is an example of how others can adopt the concerns of one or two people in a focus group.

Pictures were shown of different types of microphone for voice input (Figure 5.2). One of these was a microphone on a stand. Interestingly, participants thought this was too flimsy and could be stolen or broken off. This was a good example of a realistic image giving the wrong impression of how a future input device might look. If a less detailed sketch had been used, this might have been able to imply a more robust device which participants would have found more acceptable.

Figure 5.2 Picture used to illustrate a possible voice input device to participants

Whilst younger groups thought that a speech-based machine would cause greater problems for elderly people, the older focus groups seemed no more or less concerned than other groups. This highlights the need to be cautious about participants expressing views on behalf of other user groups. It is perhaps not surprising that those groups who had impairments seemed less accepting of the *status quo* regarding bank machine technology and were keen to discuss new ways of making them accessible, with or without the use of speech. Part of this was perhaps due to the fact that many had become technically sophisticated by being exposed to assistive technology generally. The sessions with disabled groups helped to formulate guidelines for supporting focus groups for participants with disabilities (Maguire *et al.*, 1998. See Appendix 1 to this chapter and Chapter 7).

It was found that if a group session is to be videotaped, it is important to get agreement from all participants beforehand. If not, this can lead to difficulties if one or more of the group does not wish to be taped. When this happened within one of the group sessions, the solution adopted (though not ideal) was to have one participant located off camera.

When taking video equipment to different locations, there will be varying amounts of room available to set up camera equipment. Often the camera needs to be set quite a distance back in order to get a wide enough angle to fit the whole group into view. Whilst it is desirable to have the group sitting around a table facing one another, if space is very limited, it is possible to place users in a few rows in order to get them all onto camera. If possible, premises should be inspected beforehand and camera equipment should be set up well in advance of the participants' arrival.

It was also found, during the sessions, to be important not to slow down or disrupt the discussion with note writing. It is thus preferable to have a second person as note-taker who may also, after each part of the discussion, reflect back the summary views of the group to check that they have captured them correctly. Alternatively, the moderator can take notes from the video recording after the session. Ede (1999) mentions a technique whereby she encourages one person to describe their day, and for others to record the speaker's activities onto note cards so that 'I can concentrate on asking good questions. Then I help the group sort and discuss the note cards'. The method for sorting the cards can vary according to what the researcher is looking for.

5.6.2 Case Study – Human Identification Using Biometrics

Description and purpose

A focus group study was carried out to investigate the attitudes of members of the public towards the use of biometric technology for personal verification using human characteristics such as the voice, face, finger or iris (Maguire and Graham, 1998). This would be carried out to confirm a person's identity before performing financial transactions, such as when using a bank machine, credit card or debit card. The purpose of the study was to examine different biometric solutions in terms of reliability, speed/convenience, security and intrusiveness, and to determine barriers to their acceptance.

Structure of groups

The study was based on seven groups drawn from members of the public, each containing between seven and nine individuals. It was considered useful to split the groups up into three age categories (18-29, 30-55 and 56 years and over) and for each category to have two groups of people: *acceptors of technology* and *resistors to technology*. The recruitment questionnaire presented in Appendix 2 of this chapter shows the questions used to categorise people according to technology acceptance or resistance. A special group of people who had been victims of credit card fraud was also assembled. This made seven groups in all.

Four of the groups were run in Loughborough (a town in the Midlands area of England) whilst the remaining three were run at company premises in London. This gave a balance between people living in a small town or its surrounding villages and those living in a major city. Figure 5.3a shows one of the sessions taking place in Loughborough. Figure 5.3b shows a room layout diagram used to support video analysis.

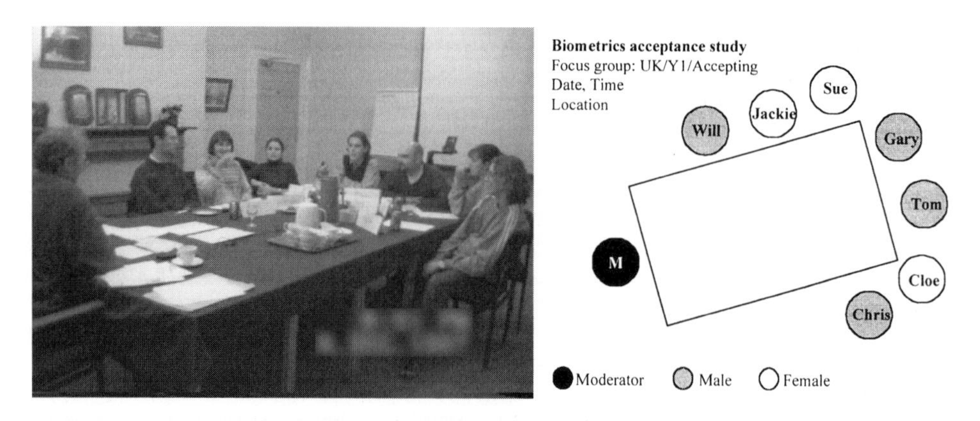

Figure 5.3a Younger, technology-accepting group, discussing biometric techniques; **b** Diagrammatic representation of group used to support video analysis

Session format

Following a session introduction, each focus group covered the following topics:

- experiences with bank cards, PINS (Personal Identification Numbers) and passwords for identification (in different situations) and general views on banking security;
- comments on telephone banking and purchasing over the Internet;
- discussion of recognition designs based on personal characteristics, including fingerprint, hand shape and size (hand geometry), face recognition, iris scanning, voice recognition and signature recognition;
- advantages or disadvantages of each technique in terms of reliability/security, speed/convenience, intrusiveness and the possibility of each technique working in a public setting (e.g. at a cash point) or in private (e.g. home banking or shopping through TV or computer);
- acceptability of biometric information being stored in the bank's central computer or only on their own card;
- acceptance and practicalities of enrolment process;
- ideal solution using the different methods discussed for a bank machine or cash point and for home shopping via a TV or computer.

Results

The focus group sessions revealed that the use of a card and PIN to access a bank machine is well established and widely accepted by bank customers. Thus many of the participants did not see the need to replace the PIN from a security point of view.

In comparing different biometric techniques, fingerprint recognition seemed to be the most acceptable method to the groups as a whole. Participants recognised that fingerprints are unique and the action of pressing a finger onto a scanning pad is convenient and could be a useful alternative when a person forgets their PIN. Generally, participants were convinced or open-minded about iris scanning although health risks were a major concern. Methods such as voice recognition and signature recognition were not regarded as reliable or secure but have the advantage, if employed, that they are seen as impersonal and non-intrusive. Face recognition was also thought to be non-intrusive (although it is camera based) but participants reserved judgment about its reliability. Voice, signature and face recognition methods were regarded favourably by those resistant to technology as they employ techniques that human beings use. Hand geometry was seen as unreliable as hands appear to the public as too similar but may be regarded as a useful backup recognition check.

It was interesting that the various focus groups had similar reactions to the different biometric techniques. One difference was that acceptors of technology were more able to see how different biometric methods might work, yet at the same time could see ways in which criminals might get around them. There was some indication that younger groups were more willing to accept iris scanning as a reliable and safe technique. Finally, the older groups were less able to comprehend the details of some of the different biometric methods and related the processes more to human capabilities. Thus the older technology-resistant group was, for example, quite positive about voice identification based on the human ability to recognise a voice.

Several participants felt they could not express clear opinions about aspects of these methods without seeing them in practice. Clearly demonstrations of techniques would be needed before people started to accept such methods and began to trust them.

One resulting idea was that if it is to be effective, the biometric method should be as unobtrusive as possible and work whilst a transaction is taking place. For example: taking a fingerprint whilst typing in a PIN; identifying a voice whilst it requests over the phone the service required; taking a print whilst a person has the phone up to their ear. Automatic checks may also go alongside human checks for maximum effect, for example, a shop assistant checking the photograph on a credit card whilst either a fingerprint is being taken or a PIN number is being typed in.

Comments on method

Many of the focus group comments demonstrated a lack of technical knowledge about how certain technologies operate. It could be argued that if the participants fully appreciated that iris scanning was simply a camera that did not flash in front of the user's eyes, they would be more accepting of the technology. Similarly, if they knew that an iris scanner could not be misled with a picture because it can detect the difference between a real eye and a picture, again they may have reacted differently. Also, if those participants were aware that the image of an eye could be digitised and compared with a stored image in a very short time, they may have had fewer doubts on efficiency grounds. Clearly a lack of knowledge about methods restricted acceptance but seeing or hearing more about them increased user acceptance. A BBC *Tomorrow's World* (future technology) TV programme, which later demonstrated the technology in an operational bank machine, received a positive response from the customers interviewed. This raises the question of whether the study should focus on initial user reactions or more informed reactions. Of course the answer to this will depend on the objectives of the study. The issue also illustrates the importance of the moderator having a good knowledge about the relevant domain of the subject being discussed.

In general it can be seen that well-established methods were preferred, for example, finger print recognition. Less intrusive methods were also preferred in principle, for example, voice verification and handwriting as they were seen as impersonal and non-intrusive. This highlighted a possible question that could have been posed to participants, i.e. 'If all of the different methods worked reliably and efficiently, which would you prefer?'. This may have been a way of overcoming their lack of knowledge about particular advanced technologies.

It was also found that setting up contrasting user groups, in terms of age and acceptance of technology, did provide a wider span of opinions, even though the differences were not great. Those who had more knowledge of technology were able to talk through the issues more thoroughly.

5.6.3 Case Study – Trader Bank Terminals

Description and purpose

A study was carried out at the HUSAT Research Institute in Loughborough to investigate the views of small traders on current banking services and on possible facilities to improve the service. In particular, the study looked at the idea of banks providing a 24-hour service to traders through self-service devices in foyers. The 24-hour bank could include:

- standard ATM or bank machine services;
- coin-dispenser;
- cash deposit machine;
- link to a bank service centre.

The study also looked at potential environments for such services and different ways of preparing money for depositing at the bank. It also investigated traders' views on the use of credit cards and charge cards in the future.

Structure of groups

Three groups of traders were recruited and discussion sessions were held. Each session lasted for about one and a half to two hours and was videotaped. Each group contained a mixture of retailers and traders (see Figure 5.4). These included managers of a public house, menswear shop, general store, off-licence, hairdressing salon, china shop, bread shop and restaurant, fresh produce and tobacco shop, gift shop, charity shop, and market traders on fashion and hosiery stalls.

Session format

Each focus group session discussed the following topics:

- the process of preparing the 'takings' (money received) for deposit at the bank;
- problems with forgeries, stolen cards and bounced cheques;
- contacts with banks, including bank charges and quality of service;
- the idea of an automated bank foyer allowing 24-hour banking – an outer area containing a cash dispenser, normal ATM, and cash deposit machine – an inner area containing a chair and terminal providing a remote telephone link to access bank services;
- location and style of an automated bank;
- the future of cash and credit cards.

Results

The discussion groups demonstrated that traders were interested in the banking services they receive and are open to ideas for improved services. Since all traders seemed sensitive to the level of bank charges, they are more likely to be encouraged by automated facilities if they get some financial benefit from them. Traders were also likely to be keener to offer credit card facilities to customers if they were able to offer traders lower charges for using them.

Figure 5.4a Trader group with facilitator and observer (the client) on the right; **b** Stimulus to demonstrate concept of 24-hour self-service foyer

The idea of the automated bank foyer was of interest to most traders although they felt it would need to be secure. Traders were particularly worried about visiting the foyer at night. Where traders are outside the town centre, an automated foyer could be of particular value. If the location could be approached in a car this would be beneficial. A brick fronted building was also preferred and with good lighting. Within the foyer, the provision of a change machine was thought to be more useful than a deposit machine. The main problem with the latter was the uncertainty that the cash would be accepted, and the lack of ability to argue if the money was miscounted.

Traders were cautious about the idea of remote links to a bank service centre. It was thought that this would be too impersonal but could be useful if human bank tellers manned the service.

Comments on method

The focus group sessions worked well and helped to reveal the main concerns about banking services and financial matters, as well as providing feedback on the ideas presented to them. This was helpful feedback for the client and gave them an idea of what the users were thinking. It was also found that the use of sketches to show new concepts rather than realistic pictures helped people to consider the ideas without being distracted by particular details.

5.7 GROUP FOCUS: PROTOTYPE REVIEW

Focus group sessions may also be conducted to provide participant reactions to a prototype system. There are, however, limitations in this type of activity. Focus groups may be able to provide general reactions to a system presented to them, but cannot reliably comment on the usability of a system without using it themselves (Nielsen, 1997). An alternative is to ask users to use a system in their own time and then to come together to comment upon it as a group. The following case study describes a focus group evaluation of a prototype digital TV service.

5.7.1 Case Study – Electronic Programme Guide

Description and purpose

A series of studies were carried out to evaluate a new electronic programme guide (EPG) for a digital TV service (Maguire *et al.*, 1999). The context of use analysis highlighted the following characteristics of usage:

- viewing at a distance;
- limited space to display information;
- interaction via a handset;
- integrated with TV programmes;
- used on a casual basis without use of a manual and in relaxed mode.

The system included a range of features. A 'Now and Next' facility showed in a small window what programme is showing on the current channel and what is coming up. The user could filter the large number of channels by type (e.g. sport, movies, children's, documentaries) or select to see programmes by subject. Programme details were presented either in list format (as in a newspaper) or grid format. Reminder, recording and parental control facilities were also provided.

A series of three discussion group sessions were held as part of digital TV trials involving members of the public. Each group was presented with demonstrations of two digital TV services, and invited to compare and comment on them. Prior to this, a series of user tests had been carried out to test the usability of the two services (see Figure 5.5) – these also helped inform the design of questions for the focus groups. The participants were recruited through a local recruitment agency according to specific characteristics.

Figure 5.5 User testing in advance of focus groups helped define issues for exploration during the discussions

Structure of groups

The three groups consisted of people who had either standard satellite or cable TV at home, so were used to concepts such as a large number of channels and paying for programmes. They were also asked to rate their level of experience of computers generally, personal computer, Teletext, Fast-text keys, the World Wide Web, and of satellite or cable TV. The composition of the three groups was as follows:

Group 1: Mainly parents of teenage children. They had the highest level of IT experience.

Group 2: Mothers between 25 and 64. They had the lowest level of experience of IT but were experienced with satellite or cable TV.

Group 3: Parents with children of 12 years and under, and had a medium level of IT experience but were experienced with satellite or cable TV.

Session format

Each focus group session lasted for two hours. The discussion covered different parts of the system and participants' general attitudes towards each system as a whole. Each session was informal with group members being encouraged to comment freely on both systems. Interviews were carried out with 13 families (including single and two parent families, and retired couples) in their homes.

Results

The main issues raised concerning the electronic programme guide (EPG) within the focus group sessions were as follows:

- clarity of layout and apparent simplicity of use;
- text for distant viewing;
- preferred style of layout of programme schedules, for example, grid versus list;
- simplicity of the parental control options;
- appeal of EPGs to men and women;
- features that were of particular value or not needed;
- one EPG appeared to have many functions, which might not all be needed;
- functions that are particularly attractive are picture in picture to watch two channels simultaneously, reminder to watch programme, list of favourite channels, and programme search.

Comments on method

The group sessions showed that users could provide feedback on the facilities available within a prototype and could discuss their needs in relation to them. However, they were less able to comment in detail on the usability of the user interface itself. They were able to provide comments on first impressions, such as 'this screen looks cluttered' and 'I would not know what to select on this screen to access set-up', but could not give clear feedback on whether they could use the system themselves.

The recruitment process (via a market research agency) succeeded well, but illustrated the compromises that sometimes have to be made. The original idea was to base the study on digital TV users. As the number of digital subscribers in the UK was small at the time of the study, this proved difficult to achieve in a short time and so the groups were based mainly on standard cable and satellite TV users.

It was found that the different compositions of each group did reveal different perspectives. For example, Group 1 (parents of teenage children) appreciated the increased convenience of being able to use the on-screen guide to set up recordings and reminders easily. They also liked the ability to filter and reduce the wide range of channels and programmes available by choosing them by subject category. The inclusion of a younger, digital user within Group 1, while differing from the characteristics of the others in the group, promoted some interesting discussion between him as a strong acceptor of digital services whilst the others were more sceptical of them. Group 2 (all mothers) were concerned about the cost of purchasing films and events, and the ability to control children's viewing.

In one of the discussion groups, there was a difference of opinion over certain features such as the 'Now and Next' window. This facility presented the name of the programme showing currently on each channel, and the name of the programme following it. The 'Now and Next' window appeared for a few moments when the user changed channel and then disappeared after a short time-out period, when the user stopped interacting with it. Whilst some users found this feature annoying, others liked it. It transpired in the discussions that several of those preferring the feature had already had experience of it on their own or on their friends' TV service. This highlighted the importance of understanding the background experience of users in order to be able to interpret their opinions. The participants' experience and preferred ways of working with their current product (e.g. the ability to type in channel numbers as well as selecting them) can also help to suggest facilities that the current demonstrations may not have.

One facility discussed was a programme planner that allowed viewers to select programmes and place them into an on-screen diary, which would then remind them to watch the programmes. This concept of a programme planner was thought to be a good idea generally. However, Group 3 participants (parents of young children) stated that they would not need them personally although 'their parents would love them', or that they would be of more interest to teenagers. This reaction against planning TV viewing may have been due partly to participants not wishing to appear as though they watched a lot of TV or that they depended on it.

Providing a demonstration of the digital service allowed participants to give their first impressions and to comment on the overall style of the user interface. The group setting also mirrored group watching of programmes so people were able to comment on whether they could read screen information on the screen at a distance. However, the demonstrations of the services did not highlight many of the interaction problems that were apparent during the individual user trials. The groups commented generally on whether they liked the layout of a screen, whether the functions looked useful or otherwise, whether the interface looked too complex. Participants were more accepting of interaction features (e.g. setting up a list of favourite channels) that individuals found quite complex when they tried to use them in the trials. Thus system demonstrations to groups should be supplemented with hands-on user trials to assess usability issues.

5.8 SUMMARY AND CONCLUSIONS

A number of general findings and recommendations arise from the focus group studies described. These are supplemented with practical guidelines and techniques for running focus groups presented in the appendices to this chapter.

Focus group sessions seem to fit in well within the early stages of the development cycle, identifying user needs, responding to new design concepts and reviewing user experience of, and giving initial impressions of, prototypes. They enable the design team to understand user perception of the product *vision*, of the kind of uses the product could be put to, and the image the product should have. They can also bring to light poor or annoying features of a product that had not been suspected during the design process.

Focus groups are often used when it is planned to bring new technology into an organisation in order to find out how the employees will react to the technology and envisage it being used. But it is usual in focus group work that the group itself undergoes a process of change as a result of meeting and discussing the issues. Thus focus groups may need to be repeated to capture the evolving ideas.

Focus groups are normally held in a neutral location although they can be run successfully in a household to interview family groups. It is important that the goals and procedure for such meetings are carefully explained to each group beforehand, to be flexible in organising and running them, and to avoid bringing too much recording equipment into the home.

It may be appropriate to give the focus group participants a list of issues in advance so they have time to think through them beforehand. It was found in the family study that creative younger members of the family liked to suggest new ideas if told something about the aims of the group meeting in advance. This is perhaps generally true of brainstorming sessions and, to be creative, the group may need thinking time before or during the meeting.

The author has been invited to run several focus group studies looking at new design concepts to assess consumer reactions. The way in which the concepts are presented is very important. Presenting a concept verbally and relying on the participants' perception of it is appropriate if the physical form of the concept is undecided and flexible. However, where the form of the technology behind the idea is defined, a demonstration, video or set of pictures will help to show clearly what the concept is about. Minor features of the representation can bias the focus group participants in their opinions, so it is often helpful to simply sketch the ideas without too much detail to avoid colouring the participants' judgment. Table 5.1 summarises the advantages and disadvantages of using demonstrations in focus groups.

Table 5.1 Advantages and disadvantages of demonstrations in focus groups

Advantages	Disadvantages
Allows users to understand the design concept being discussed more clearly.	May raise unrealistic expectations of the capability of the future system.
Satisfies possible concerns about safety, and security for some types of system.	Users may take explanations of the usability of the system for granted.
	Demonstration may lack features (e.g. broadcast sound) which may influence usage.

Having contrasting groups with certain characteristics, for example, age, IT experience or technology awareness or acceptance, is a good way to gauge the spread of opinions across the user population. In order to get a distinct view it is generally better to have homogeneity within groups rather than mixing people with different characteristics across groups.

Like most usability methods, focus groups are best used in conjunction with other methods. Administering a survey before focus group sessions are run can help to establish the basic user issues relating to the topic (as well gathering some quantitative data) before investigating them in more detail within the group sessions. Focus groups considering a product idea may also be complemented with user test sessions to investigate the usability aspects of an early prototype. Focus group sessions where users watch a system being demonstrated should not be relied upon to assess usability.

During focus group discussion, the aim is to try and reach a consensus on each issue to clarify or prioritise user requirements for a system. However, it is important to avoid forcing consensus on the group (which may be determined by opinion leaders in a group) so that the outcome on a particular issue may be a split view. Final views can be recorded by asking participants to write them down following each section of the meeting or to invite them to hold up voting cards. Alternatively, a questionnaire may be distributed at the end of a session to record individual views, possibly inviting participants to list the main features they would like on a particular system.

It is important that a focus group is run smoothly to keep it within the allotted time and to maintain participant interest. The moderator should avoid taking lengthy notes during the meeting, which may break up the flow. Thus a second researcher should take notes or the meeting should be recorded for later analysis.

The moderator should be prepared to allow groups to voice concerns or requirements beyond those topics listed in the agenda if they have a bearing on the main topics being discussed. Such additional comments may help to generate new requirements or even change the focus of the new system concept.

People often find focus group sessions stimulating and often think about the issues for several hours or days after the meeting. As a result, they sometimes have further ideas having thought through the issues at greater length. It is important that a means be provided for these additional thoughts to be captured by providing contact details for the facilitator or by administering a follow-up questionnaire.

5.9 REFERENCES AND RELATED READINGS

Caplan, S., 1990, Using focus group methodology for ergonomic design. *Ergonomics*, **33** (5), pp. 527-533.

Cross, N., 1989, *Engineering design methods*, (Chichester: Wiley), pp. 37-38.

Derbyshire, D., 1999, Cash machine phobia. *Daily Mail*, 14 July, p. 31.

Ede, M., 1999, *Focus Groups to Study Work Practice*. www.useit.com/paper/meghan_focus groups.html (accessed 13 December 1999).

ISO, 1999, ISO 13407: *Human-centred design processes for interactive systems*. (Geneva: International Standards Organisation. Also available from the British Standards Institute, London.)

Henderson, M. and Hawkes, N., 2001, Nine angry men are more than enough. *The Times*, 4 September 2001, p. 7.

Hone, K.S., Graham, R., Maguire, M.C., Baber, C. and Johnson, G.I., 1998, Speech technology for Automatic Teller Machines: an investigation of user attitude and performance. *Ergonomics*, **41** (7), pp. 962-981, ISSN 0014-0139.

Jones, J.C., 1980, *Design methods: seeds of human futures*, (Chichester: Wiley). ISBN 0-471-44790-0.

Kuhn, K., 2000, Problems and benefits of requirements gathering with focus groups: a case study. *International journal of human-computer interaction*, **12** (3 & 4), pp. 309-325.

Macaulay L.A., 1996, *Requirements Engineering*, (Berlin: Springer Verlag Series on Applied Computing), ISBN 3-540-76006-7.

Maguire, M., Heim, J. Endestad, T. Skjetne J.H. and Vereker, N., 1998, *Requirements specification and evaluation for user groups with special needs, WP6 – Deliverable D6.2, Version 2.0*, (HUSAT Research Institute, Loughborough), 25 August 1998.

Maguire, M., 1998, *Report of trader discussion groups*, (HUSAT Research Institute, Loughborough), 10 April 1998.

Maguire, M. and Graham, R., 1998, *Human aspects of biometric methods for verification: focus group report*, (HUSAT Research Institute, Loughborough), 9 March 1998.

Maguire, M., 1999, *Report on study of the management of domestic finances by family groups*, (HUSAT Research Institute, Loughborough), 7 May 1999.

Maguire, M.C., Graham, R. and Hirst, S.J., 1999, *Digital TV electronic programme guide comparative study – Phase 3 Report on user trials and focus groups*, (HUSAT Research Institute, Loughborough), 12 March 1999.

McClelland, I.L. and Brigham, F.R., 1990, Marketing ergonomics – how should ergonomics be packaged? *Ergonomics*, **33** (5), pp. 519–526.

Nielsen, J., 1993, *Usability Engineering*, (London: Academic Press – AP Professional), ISBN 0-12-518405-0.

Nielsen, J., 1997, The use and misuse of focus groups. In *The Focus Group: A Strategic Guide to Organizing, Conducting and Analyzing the Focus Group Interview* (2nd edition), Jane Farley (ed.) (Templeton: Probus Publishing). See also www.useit.com/papers (accessed 12 May 1999).

Nielsen, J., 1999, *Voodoo usability.* www.useit.com/alertbox/991212.html (accessed 12 May 1999).

Osborn, A.F., 1963, *Applied imagination*, (New York: Scribeners and Sons).

O'Donnell, P.J., Scobie, G. and Baxter, I., 1991, The use of focus groups as an evaluation technique in HCI. In *People and Computers VI.* Diaper, D. and Hammond N. (eds), (Cambridge, UK: Cambridge University Press), pp. 211-224.

Poulson, D., Ashby, M. and Richardson, S. (eds), 1996, *USERfit A practical handbook on user-centred design for assistive technology*, Handbook produced within the European Commission TIDE programme USER project, (HUSAT Research Institute, Loughborough).

Stewart, D.W. and Shamdasani, P.N., 1990, *Focus groups: Theory and practice. Applied Social Research Methods Series*, **20**, (Newbury Park, CA: Sage Publications).

Acknowledgements

The author is grateful to his partners within the focus group studies reported including his former HUSAT colleague, Robert Graham, Alexandra Trabak and Paula Lynch of NCR Knowledge Lab, Michael McNamara of NCR Financial Systems Ltd and Marie Kelman of NTL.

APPENDIX 1: GUIDANCE ON RUNNING FOCUS GROUPS WITH PHYSICALLY AND VISUALLY IMPAIRED PARTICIPANTS

For dexterity and mobility impaired groups:

Arrangement for transport to and from the meeting should be considered to be the organiser's responsibility.

The accessibility of the location must be considered in detail before the discussion is arranged. It is not only the room itself that has to be accessible, but also the immediate environment, for example, the toilets, the dining room and the table that the participants sit around.

Wheelchair access to the building where the interview is to be held, and its facilities, should be provided. Even if wheelchair users are not involved it is preferable that visitors do not need to climb flights of stairs to reach the group discussion area.

Physically impaired people may rely on friends and relations to transport them to and from the location for the discussion. It is important then that the discussion finishes on time so as not to inconvenience either the discussion group members or their helpers. Provision should be made for those assisting to wait, if they prefer, with extra seating or a comfortable waiting room whilst the discussion proceeds.

Ensure that the physical disability of each focus group member is properly recorded, for example, 'disabled so uses wheelchair but can travel a short way on crutches', 'cerebral palsy so cannot use right arm'.

For visually impaired groups:

Users who are visually impaired will need to get familiar with the location where the discussion is held. The focus group organisers should be prepared to spend some time showing the participants the layout of the room, how to get to the toilet etc.

The participants should also be asked if they want guiding or some other kind of help. Remember that a visually impaired participant will want to hold the guide's arm rather than the guide holding theirs.

Ensure that the lighting conditions are adequate for the participants' vision. Ask in advance what requirements the participants have. Some people will need high intensities, whilst others prefer dim light.

Group members with normal sight will often use visual cues to indicate their wish to speak and may not realise that this is of no use to people who are blind and/or visually impaired. It is important that only one person is allowed to speak at any time, and that the discussion leader manages this. It may also be appropriate to have a simple rule to indicate when a person wishes to speak, i.e. raising their hand for attention.

If name cards are to be displayed for each discussion group member, prepare them beforehand rather than relying on the participants themselves to write them. Provide space and perhaps water for guide dogs accompanying the group members. If the discussion relies on visual presentation of written materials, participants with low vision usually need an enlarged font and good contrast between text and paper. For users who are visually impaired, 16 point font or larger should be used. Try to present all information verbally as well as visually.

Although persons with visual impairment may gain little benefit from the use of visual aids such as overheads or whiteboards, other participants in the group may want to use such aids. The discussion leader should discuss this with the group and find a solution that is satisfactory to all participants. For example, visual text may also be read out to those who are visually impaired.

People with visual impairment may rely on friends and relations to transport them to and from the location for the discussion. Therefore make sure that the discussion finishes on time so as not to inconvenience either the discussion group members or their helpers. Provide comfortable facilities for helpers to wait, if they prefer, while the focus group session takes place.

The group leader should wear bright clothes to make sure that all the participants can see him or her.

Where pictures or demonstrations are to be shown, the presenter should be prepared to describe them as well as show them, so that all participants (including those with visual impairments) are able to appreciate what is being shown.

Ensure that the visual disability of each focus group member is properly recorded, for example, 'registered blind with retinitis pigmentosa', 'good peripheral vision but decreased central vision'.

APPENDIX 2 – TECHNOLOGY ACCEPTANCE QUESTIONNAIRE

(Used within case study on biometrics for verification)

Please answer the following questions concerning your attitude to technology:

(1) Do you like to use the full range of functions on a home product such as a telephone, music system or washing machine? *(If yes, score 1 point)* Yes ❑ No ❑

(2) Do you prefer to use an automatic bank machine rather than seeing a bank clerk for basic bank services? *(If yes, score 1 point)* Yes ❑ No ❑

(3) Do you enjoy exploring the possibilities of new technology? *(If yes, score 1 point)* Yes ❑ No ❑

(4) Do you find it difficult leaving messages on an answering machine? *(If no, score 1 point)* Yes ❑ No ❑

(5) Do you think that computers create more problems than they solve? *(If no, score 1 point)* Yes ❑ No ❑

(6) Is access to the Internet important to you? *(If yes, score 1 point)* Yes ❑ No ❑

(7) Do you prefer to use a simple camera rather than an advanced one? *(If no, score 1 point)* Yes ❑ No ❑

(8) Would you like to see more services offered through a bank machine, such as requesting loans or obtaining financial advice? *(If yes, score 1 point)* Yes ❑ No ❑

(9) Do you use a mobile phone regularly? *(If no, score 1 point)* Yes ❑ No ❑

(10) Do you like to use innovative technology as opposed to tried and tested technology? *(If yes, score 1 point)* Yes ❑ No ❑

(11) Do you regularly program a video recorder? *(If yes, score 1 point)* Yes ❑ No ❑

(12) Do you think that we rely too much on computers? *(If no, score 1 point)* Yes ❑ No ❑

Those scoring 6 or less are allocated to a 'technology resistant' group.
Those scoring 7 or more are allocated to a 'technology accepting' group.

CHAPTER 6

Focus Groups in Health and Safety Research

Roger Haslam

6.1 INTRODUCTION

This chapter describes the use of focus groups in health and safety research, in both occupational and domestic settings. Reasons for using focus groups are discussed, along with the benefits and difficulties of adopting the approach. Three case studies are presented: (1) stair safety amongst older people, (2) causes of accidents in the construction industry and (3) effects of medication prescribed for anxiety and depression on work performance. Although wide-ranging topics, the three case studies illustrate different approaches to the recruitment of subjects, analysis of data and presentation of findings. Lessons learnt from our use of the method are discussed. The chapter concludes by emphasising the value of focus groups as an adjunct to other more established methods of collecting subjective data from human study populations.

Focus groups have been used by the Health & Safety Ergonomics Unit (HSEU), Loughborough University, UK, for a number of years. Our use has ranged from 'quick and dirty' to circumstances where we have attempted to be more formal in their application (e.g. Morgan and Krueger, 1998). The work of the HSEU encompasses health and safety issues in both work and non-occupational settings, mostly carried out as large-scale research projects running over 1-2 years, within a university context.[1] This latter point is important, as although most of our work is funded by organisations interested in practical matters (e.g. Department of Trade and Industry, Health and Safety Executive, Royal Mail), there is a need for us to have traditional academic outputs from our research; dissemination through refereed journals in particular. Notwithstanding this requirement, we have found focus groups to provide a highly effective, efficient way of investigating human factors aspects of health and safety problems.

Our research has used focus groups as a means of identifying experiences, attitudes and beliefs within a study population, having the advantage that the group discussion serves as a prompt to individual participants, encouraging further thought and contribution. The focus group methodology also allows the opportunity for peer commentary on opinions expressed by others. Limitations are that: views expressed may not accurately represent those of the population from which participants have been recruited; careful moderation is required to ensure opinions of more vocal participants do not dominate those of less forthcoming individuals; and data obtained may indicate the way a situation is perceived rather than how it actually is. Sometimes what participants say is factually correct, but on other occasions they may be presenting beliefs or opinions which may be inaccurate or with which others may disagree. It is important to appreciate that in many ergonomics situations, however, it is valuable to know what users think, whether or not they are accurate in what they believe (Haslam, 2002).

---

[1] See the HSEU website for further details at http://humsci.lboro.ac.uk/hseu

The data yielded by focus groups are of course mostly descriptive and qualitative, with the method of limited value in providing quantification. But sometimes quantification is a luxury clients are not particularly interested in. On occasions, we have administered questionnaires to focus group participants, to count up experiences or views, but the small number of individuals involved usually means confidence in terms of representativeness is low.

For most work in the HSEU, although not exclusively, focus groups have been used in combination with other methods, such as interview/questionnaire surveys or laboratory studies. The technique has usually been used in the early stages of a project to inform later phases of the research and to provide an efficient route to increasing researchers' knowledge of an issue. On this latter point, focus groups provide a useful induction to researchers new to a research topic. Elsewhere, we have used focus groups to check conclusions or recommendations derived from other research methods, acting as a sounding board for the acceptability of these.

Focus groups can be relatively quick to conduct. Preparation and analysis, on the other hand, can be very time-consuming. Subject recruitment is a particular issue and may sometimes require substantial effort. Affecting this is the ease with which the study group can be accessed and their ability and willingness to convene at a given place and time. In focus group research with older people, Haslam *et al.* (2001) found it easy to recruit physically active participants but more difficult to gain contributions from those with reduced mobility. It was less easy to inform subjects who do not get out much about the research and it was then more difficult for them to travel to the venue of a focus group meeting. A project with the construction industry (Hide *et al.*, 2001) encountered problems making contact with certain study groups to seek their participation, due to their being geographically dispersed and with limited avenues for publicising the study to potential subjects. Hastings *et al.* (2002) had to devote considerable resources to recruiting participants for a study examining the effects of medication prescribed for anxiety and depression on work performance. In this instance, the personal nature and sensitivity of the topic may lead sufferers to wish to keep their experiences to themselves.

Methods we have used to enlist participants for different studies have included:

- addressing meetings or other gatherings which attract target groups;
- notices (posters) where target groups convene;
- newspaper articles and advertisements;
- requests for assistance on appropriate radio programmes;
- dissemination of study details through employer organisations and trade unions;
- direct contact with and through employers.

It is most efficient in terms of project resources to communicate with target participants directly (where this is possible). Recruitment through newspaper and other media usually generates interest, but requires significant staff effort to deal with the large number of telephone, e-mail and written enquiries that may arise.

Analysis of the information yielded by focus groups can also take a considerable amount of time, depending on how formal this is. Transcription of audio recordings is often the starting point for analysis, with one hour of recording typically taking ten hours to transcribe. An alternative to full transcription can be partial extraction of key points. This needs to be done by the researchers, however, as it requires understanding of the research issues under investigation. Our approach to further analysis is then to identify themes and categories arising within the focus group discussions. To improve the objectivity of this, one researcher might undertake an initial analysis, with one or more other researchers then providing an independent check of this. The reliability (repeatability) of the analysis is important to the scientific integrity of the conclusions.

Three case studies follow illustrating our use of focus groups addressing different research questions. Although diverse in the topics addressed, the case studies indicate the different ways in which we have used focus groups, the findings that arise and varying styles adopted for analysing and reporting results.

6.2 STAIR SAFETY FOR OLDER PEOPLE IN THE HOME

This research, undertaken for the Department of Trade and Industry, investigated the interaction between the behaviour of older people, features of the environment and the risk of falling on stairs in the home. The study was commissioned to inform the UK government '*Avoiding slips, trips and broken hips*' campaign. A serious of focus groups were undertaken at the outset of the project, intended to provide initial insight into issues surrounding the problem (Haslam *et al.*, 2001). The findings from this stage of the project were used to guide the design of a survey entailing visits to the homes of 150 older people, where the occupants were interviewed and a stair safety audit undertaken (Hill *et al.*, 2000). Central to the study was that the views and experiences of older people should be to the fore. Other research in the area has frequently been undertaken in a clinical context (e.g. American Geriatrics Society, 2001) and has been criticised by this author for treating falls as predominantly a medical matter.

6.2.1 Participants

Sampling was on a convenience basis, with participation criteria being an age of 65 years or over, with individuals living independently in their own homes. Initially, participants were recruited from existing subject lists held at Loughborough University, compiled for previous research on the risk of osteoporotic fracture in women (Brooke-Wavell *et al.*, 1995). These lists were originally drawn up through letters sent by GPs to patients on their practice lists, aged 60-70 at that time, and in good health. The lists were subsequently extended by Brooke-Wavell *et al.* through inclusion of acquaintances of the existing cohort and members recruited through community groups (over 50s recreation classes, Women's Institute, University of the Third Age and local voluntary groups). For the home falls research, other participants were obtained by displaying posters in post offices, community groups and other places where older people were considered likely to congregate. A consequence of the recruitment process was that the sample comprised more women than men. The final sample was 20 females and 4 males, with ages ranging from 65 to 79 years. This imbalance was considered acceptable given the demography of the older population, where women exceed men in numbers.

Participants formed three groups, each group meeting on a separate occasion. Numbers of participants in each group were 6 (4 females, 2 males), 7 (6 females, 1 male) and 10 (9 females, 1 male). There were 2 married couples within the sample of 24, resulting in contributions from 22 households. The composition of the groups covered a broad range of housing, dwellings varying in age from 10 to 100+ years, with subjects drawn from both urban and rural areas.

6.2.2 Procedures and Analysis

Participants were briefed both verbally and in writing about the study prior to arrival. They were informed that the discussions would consider stairs in the home and were asked to think about how and why they use their stairs, and the risk factors and safety issues that might be involved. On arrival, participants were invited to read a leaflet from the UK government '*Avoiding slips, trips and broken hips*' campaign. As well as practical suggestions for improving safety, the campaign advice covers physical fitness, diet, clothing, lighting and home maintenance. The broad content of the leaflet was thought unlikely to constrain or direct the group discussions unduly. In addition, photographs were circulated showing examples of different types of stairs. The purpose of the briefing was to encourage thought and to promote discussion.

Each meeting lasted approximately 1-2 hours and was led by a principal and assistant moderator, both members of the research team. Topics covered by the sessions were:

- circumstances in which stairs are used;
- factors affecting safety on stairs;
- personal factors leading to increased risk of falling;
- self-perceived safety on stairs;
- immediate and longer term consequences of having a fall;
- value and acceptability of preventative measures.

The focus group discussions were recorded on audio-tape, with the consent of participants, and the recordings were subsequently transcribed. Data analysis involved one of the researchers identifying concepts within the data and classifying these into appropriate categories. The other researcher working on the project then independently checked the first researcher's interpretations. An additional member of the research team undertook a further review of the analysis, referring back to the transcripts on a sample basis. In this instance, the transcription process did not record details of each speaker's characteristics (e.g. gender and age) alongside their contributions to the discussion, and thus are not noted below. Although not essential, the inclusion of such detail does provide useful context when analysing and presenting the discourse and is a practice we have adopted in more recent studies.

6.2.3 Findings

The focus groups led to a rich collection of experiences and personal accounts, and allowed the researchers to propose mechanisms that might lead to falls on stairs (Figure 6.1). Extracts from the results are given below, which illustrate the general nature of the findings and also where product design might be involved or where opportunities for product development could exist.

As with other types of accidents, falls on stairs usually involve a combination of factors, which in isolation would be less serious. As one participant recounted:

> *'I got into bed, I'd had a drink and realised that I hadn't got any meat out for Sunday, rushed downstairs and slipped. The shopping trolley was at the bottom, I wouldn't have hurt myself if it wasn't for the trolley ...I cracked my ribs. But that was my own fault, I had had a drink, and I had nothing on me feet.'*

Poor lighting can be a factor in falls and provision of artificial light needs to be adequate for use in areas with insufficient natural lighting and at night. Long-life light bulbs (compact fluorescent) were discussed in the focus groups, provoking mixed reactions. A number of respondents thought long-life bulbs might be safer than ordinary bulbs, as they need to be changed less often. Against this, others mentioned problems due to the bulbs taking longer to 'warm up' to produce their maximum illumination: '*...when you first put them on, they're very dim, aren't they? It takes quite a while to build up'.* This might actually be an advantage in some circumstances, allowing gradual light adaptation, such as when switching a light on after waking during the night.

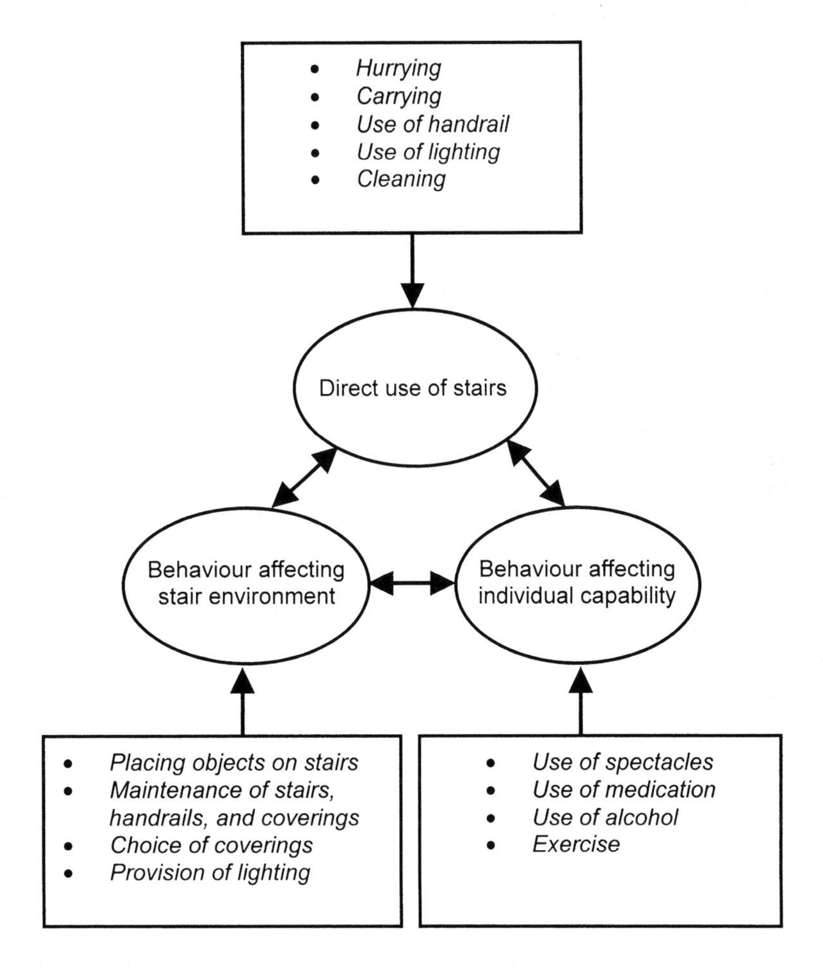

Figure 6.1 Behaviour-based risk factors for older people falling on stairs

There was unanimous recognition that leaving objects on stairs increases the chances of having a fall, with objects both reducing the usable width of the stairway and forming a hazard to slip or trip on. However, this did not prevent participants from continuing to store or temporarily place objects on their stairs. Most admitted leaving things on stairs: '*...ready for the next journey up*' or to alleviate having to carry too much at one time. The discussions suggested that items left on the stairs are usually small items, such as shopping or laundry and are put: '*...to one side on the bottom two or three steps*', '*...where people aren't going to walk*', '*...and then the next time you go up, you take them with you*'. A further source of objects on stairs can be grandchildren leaving their toys. This was mentioned as requiring special attention:

> *'I'm always careful with the grandchildren, to make sure that they don't leave things on the stairs...I always make sure that sticklebricks and things like that aren't left, as that's a major hazard I think.'*

Although there was general agreement among focus group participants that hurrying on the stairs increases the likelihood of having a fall, people appear to continue to do it. The most common reasons given for hurrying were to answer the telephone or doorbell, prompted by a concern that callers will not wait.

To make answering the telephone easier, many people had installed a telephone extension upstairs. Also, the introduction by British Telecom in the UK of a number recall system, whereby telephone users can dial '1471' to be told the telephone number of the last caller, was mentioned as alleviating the problem to some extent. However, it was suggested by a focus group member that people still rush:

> *'We're all from an age where we used to rush for the phone, surely, I mean the phone was dominant at one time...you always think that it's going to be something important. There is still that tendency to rush.'*

A further cause given for hurrying on stairs was the setting or cancelling of intruder alarms:

> *'We can isolate the upstairs fortunately, but if you forget to do that, you've not got a lot of time to race downstairs and switch it off. That can be hazardous.'*

In the group where this was discussed, two other participants mentioned that their intruder alarm leads them to hurry on the stairs or make additional stair journeys through forgetting to set it.

A number of participants reported that cleaning around the stairs, particularly using a vacuum cleaner, becomes increasingly difficult with age. Some had overcome this by using a battery operated hand-held cleaner: '*...and I have got a smaller one which I use on the stairs, which is very useful, because you can actually hold it, it's lighter*'. Other aspects of cleaning, such as accessing certain designs of window, were also thought to be dangerous.

The use of inappropriate spectacles was suggested by some focus group members as being an important risk factor in falls among older people. Bifocals were described as affecting the ability to judge distances, for example: '*...I'm forever hitting the floor with my feet, I think it's there and there's a big drop*'. Bifocal and varifocal lenses may create the need to angle the head to be able to see properly when coming down stairs, depending on the prescription. Several group members suggested this may have implications for balance:

> *'I think elderly people who have either bifocals or varifocals are at a disadvantage going downstairs, because of this necessity to get the head out of alignment with the body, in order to see properly.'*

Some participants also remarked that it is difficult to take a quick glance at anything: '*You can't make a casual glance and then know where you are, you've got to think about it*'. Some participants reported that their optician or optometrist had warned them of problems and had given advice such as: '*Don't be afraid to show your double chin*'.

In connection with failing eyesight, Figure 6.2a illustrates how choice of floor covering may affect the ability to distinguish steps. Although the focus group participants discussed the importance of having well fitted and maintained stair coverings, they were not aware that surface pattern or colour can act to camouflage step edges.

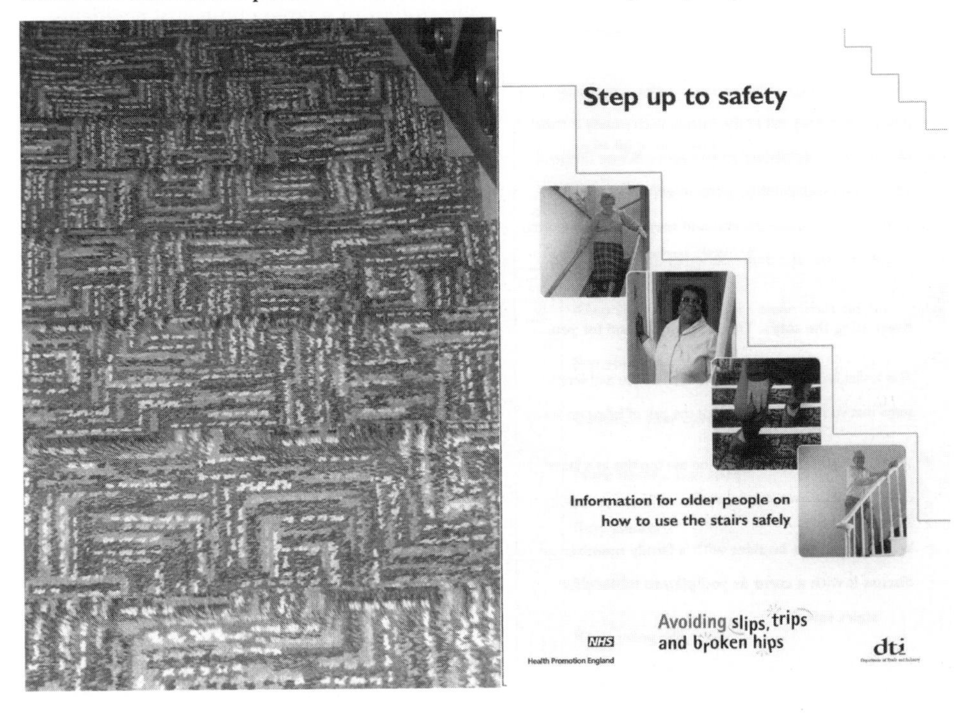

Figure 6.2a Surface patterning can camouflage step edges; **b** Stair safety guidance leaflet

6.2.4 Summary

The focus groups in this study provided good insight into factors affecting falls amongst older people on stairs in the home. Of particular value were the verbatim accounts in providing rich, illustrative material, highlighting the circumstances in which falls occur and contributory factors. This was useful to the project clients, providing material for publicity activities. It is interesting that some difficulty was encountered in finding a scientific journal interested in publishing the results from this study. A problem with many journals is that the value of qualitative data is not always recognised by those educated to believe in the pre-eminence of an experimental approach. In addition to traditional research outputs, the research also provided the basis for stair safety guidance, disseminated through the '*Avoiding slips, trips and broken hips*' campaign (Figure 6.2b).

6.3 CAUSES OF ACCIDENTS IN THE CONSTRUCTION INDUSTRY

The construction industry is one of the most dangerous employment sectors in the UK, with injury rates amongst the highest for both fatal and major injuries. Construction sites are complex, dangerous environments (Figure 6.3). Whilst previous research has provided a good understanding of the extent and pattern of accidents in construction, there is only limited objective evidence regarding the full range of contributory managerial, site and individual factors. The HSEU has been undertaking a substantive project, funded by the Health & Safety Executive (HSE), addressing this problem. Focus groups were used during the initial stages of the research (Hide *et al.*, 2001) to inform the main phase of the project, with detailed follow-up studies of a large number of accidents.

Figure 6.3 Construction sites are complex, dangerous environments

6.3.1 Method

Broad areas for discussion were identified based on research published previously concerning causal factors in construction industry accidents, namely problems occurring with:

- project concept, design and procurement;
- work organisation and management;
- task factors;
- individual factors.

Prompts were prepared to stimulate discussion for each topic, providing clarification of the sphere of interest of the study.

6.3.2 Participants

All groups had between five and seven participants, with the planned composition shown in Table 6.1. Groups one and three varied from original intentions, mostly due to the practicalities of recruiting participants. This had implications for the results, an issue discussed further below.

Table 6.1 Focus group participants

Group	Employment	Target participants
One	Client team	Client or client representative, architect, engineer (structural/civil or mechanical/electrical), financial manager, project manager or design manager and a planning supervisor
Two	Senior managers	From general and specialist contractor firms representing civil engineering, major building or the residential sectors and from small and large projects
Three	Site managers	
Four	Operatives large site	Tradesmen or general operatives
Five	Operatives small site	
Six	Safety professionals	Industrial safety professionals and construction enforcement officers
Seven	Mixed group	A mixed discipline group (trades and professionals)

6.3.3 Procedures and Analysis

Each group was scheduled to last approximately 90 minutes, a concern being to achieve a balance between allowing sufficient time for discussion and to avoid deterring busy people from participating. The two group moderators introduced the prompts prior to each of the discussions, and participants were encouraged to explore beyond the themes introduced. With each prompt, participants were asked to consider 'where does failure occur?' and 'why do accidents still happen?'. All discussions were audio-taped and an abridged transcription was made for each focus group. Analysis included an intermediate summarisation of all text into short bullet point statements per group, followed later by comparison and categorisation of all focus group information according to the discussion area headings provided. Findings were subsequently reviewed by industrialists during two separate conference gatherings as part of the validation process. The use of summary statements, as presented below, was prompted by a need to condense the volume of data generated from the groups into much more concise form. Whilst this loses some of the detail and impact of verbatim quotes, it aids dissemination where brevity is required.

6.3.4 Findings

Considerable information was generated under each heading. The following are example summary statements compiled from the discussions relevant to design practices (with 'design' considered in its broadest sense).

Inadequacies in design:

- compatibility of building items not considered – for example, compatibility of item weights with available lifting gear;
- tenders are made on the basis of design drawings, but if they are later found to be wrong, it is rarely possible to change the funding/work schedule – this can lead to short-cuts and subsequent higher risk of accidents;

- design modifications of work in progress are not comprehensively considered in the context of the whole design, leading to unforeseen 'knock on' effects;
- building designers do not consider maintenance.

An example relevant to this was where a participant reported that maintenance to windows on a particular tower block was only possible from outside, as the windows did not open inwards. Therefore, they had to put up a scaffold every time work was required – costing thousands and with the safety implications of using scaffolding. Had the windows opened inwards they would have been easy and cheap to maintain.

Inadequacies in planning:

- planners just focus on the task, not site layout issues and related aspects, such as traffic management, leading to disorganised, untidy and unsafe work places;
- 'just in time' is not considered and parts delivery and storage can exacerbate problems with layout and task area design.

Communication issues:

- quantity surveying practices can ruin good things from design;
- the industry culture is to blame everybody else for problems;
- the tender document and pre-tender health and safety plan often have a number of meaningless statements in them.

One example recalled by a participant was '*The hazards associated with this project are not beyond the competence of a capable contractor to control. If the contractor should find any hazardous material he should notify the client and the client will then give direction of what should be done*'. The participant reported, however, that frequently they are not notified by the design team as to what the project hazards are, as ought to happen under the Construction (Design and Management) Regulations 1994 (CDM) (HSC, 2001).[2]

Legislation related issues:

- the Planning Supervisor (a function required under CDM) has insufficient authority to ensure that designers account for CDM responsibilities;
- designers did not want CDM, although they are now starting to think about safety, but not health;
- people do not understand the legislation and blame the HSE;
- the CDM regulations are just a paperwork exercise and do not enhance health and safety.

Design innovation:

- the HSE are urging for innovation to improve design factors, but this responsibility has fallen onto the shoulders of contractors and not the client team;
- more things now get made offsite in factories (as it has become more difficult to get skilled trade-people on site) and this can inhibit good design.

---

[2] The CDM regulations in the UK impose explicit duties on those responsible for construction design, requiring that they seek to eliminate hazards and reduce risks to construction workers. However, the regulations have proved controversial and compliance has been patchy. The Health and Safety Executive (HSE), the government body responsible for safety policy and regulation, have recently attempted to address the problems with updated guidance (HSC, 2001).

6.3.5 Summary

It should be apparent from these brief extracts that the focus group discussions provided a provocative commentary, reflecting views held amongst those working in the construction industry. Difficulty recruiting clients and designers for the study meant that their views were lacking from the focus group data *per se* and it is evident from the opinions presented above that designers, in particular, were subject to a degree of criticism from others in the industry. This was rectified to some extent through the validation sessions, one of which was held with participants at an international conference devoted to designing for construction safety. Interestingly, the validation exercises supported the focus group data concerning problems with the design-safety interface, suggesting the issue to be more than a matter of professional rivalry. There does seem to be a lower acceptance of responsibility for health and safety amongst the construction design community than others in the industry believe should be the case.

Although informative in their own right, the findings from the focus groups provided a strong basis for the main part of this research: detailed follow up studies of a large number of accidents. The focus groups set the agenda for these accident studies, with the studies structured to provide evidence to confirm or refute assertions made in the focus groups. At the time of writing, this follow up work is still in progress.

6.4 USE OF MEDICATION IN THE WORKING POPULATION AND EFFECTS ON PERFORMANCE

The extent to which medication is used to treat anxiety and depression in the working population is unknown, although levels are thought to be considerable (Dunne *et al.*, 1986; Potter, 1990). These and other studies (e.g. Nicholson, 1990; Tilson, 1990) have highlighted the poor knowledge that exists concerning the effects of prescribed medication on work performance.

Laboratory studies have found that psychotropic medicines (those capable of affecting the mind, emotions and behaviour) used to treat anxiety and depression impair performance on a wide range of measures, including attention, vigilance, memory and motor co-ordination. However, it is not clear how these findings translate to performance in the workplace. There are problems with generalising from laboratory studies as they are often limited to testing with young, healthy subjects and minor decrements in performance on sensitive laboratory tasks may have little relevance to real world activities. It is not clear, therefore, to what extent medication may affect quality of work, productivity and safety.

Lack of treatment for psychiatric illnesses may actually be a greater problem in terms of work performance than the side effects of medication. Employees suffering with anxiety or depression are likely to experience a range of symptoms that would impair performance at work, including: tiredness, lack of motivation, poor concentration and forgetfulness, and poor timekeeping and attendance. Failure to seek or comply with treatment may arise from the stigma associated with mental ill-health or, alternatively, some people may be reluctant to take medication because of fears about side effects. This research collected new data on the use of psychotropic medication amongst the working population to improve understanding of the impact of mental health problems and the treatment for these conditions on performance and safety in the workplace (Hastings *et al.*, 2002).

6.4.1 Procedures and Analysis

This research depended solely on focus groups. The technique was considered well-suited to gathering information about day-to-day experiences of mental health problems and the impact of medication on work performance. The focus groups also afforded participants a degree of discretion as to how much personal information to reveal, more so than might have been the case in a one-to-one interview.

A total of 12 focus groups were conducted across a broad spectrum of employment sectors, with each group led by two moderators. Three groups were held with participants from anxiety and depression management courses run by clinical psychology services, with participants originating from a wide range of professions. Three groups comprised staff from Human Resources and Occupational Health and Safety departments across a range of organisations. These groups provided information regarding organisational perspectives on the mental health of employees. The remaining six focus groups were conducted with participants reporting personal experience of anxiety or depression, recruited directly. The study strategically targeted industrial sectors known to be high-risk for mental health problems (health care, education, extraction and utility industries, and manufacturing).

A major challenge for this research was the recruitment of participants. The sensitive nature of the topic and the embarrassment associated with mental ill-health were potential obstacles to this. Extensive recruitment activities were undertaken on a national and local level, involving media contact, advertising, publicity through trade unions and professional organisations.

The focus groups discussed respondents' experiences of anxiety or depression; use of prescribed medication for these disorders; compliance with treatment regimens and improvement in symptoms. The groups dealt with both the effects of mental health problems and medication treatments on work performance, including effects on absenteeism and risk of accidents, effects on relationships with colleagues and the extent to which employers support employees who are experiencing mental health problems. The groups with organisational representatives concentrated on recruitment, support and rehabilitation of employees with problems. Topical statements about anxiety and depression were used at the beginning of each session to encourage general discussion and to generate interaction between participants.

6.4.2 Findings

The focus groups were recorded and fully transcribed. The data were analysed by sorting verbatim material into thematic categories, with the categories derived and reviewed by the researchers undertaking the analysis on an iterative basis. For this project, the client specifically requested that selected vignettes be included in the report, in addition to representative quotations, giving emphasis to individual experiences (see Figure 6.4, for example).

Example results indicated that sufferers found it difficult to distinguish between symptoms of anxiety and depression and the effects of medication. Both the symptoms of anxiety and depression and medication for these conditions were reported to impair work performance. Participants described a variety of accidents and near misses that they attributed to their condition or to the side effects of medication.

> *'I had a series of falls at the time...and, in retrospect, I don't think it was the medication. I think part of my problem was the job piled up, I would go faster and*

faster and faster. So if I'd got a large queue, I'd try and deal with them really quickly because I know they're all waiting. And I don't need to wee or have a break, because this lady's waiting here. And I go faster and faster and I'm sure I would walk faster and faster. And I had a series of falls and they actually sent me for some tests because of it.'

(female working in university administration)

'I tend to make silly mistakes, some of them resulting in cuts and scratches, you know minor injuries.'

(female working in the public sector)

A male assembly line worker in a major confectionery company had been working on a production line and over the years the number of people working on the line dwindled down in numbers. Temporary staff were brought in and as this man was the longest serving member of staff on this particular line he was expected to take on a supervisory role:

'Like he was relying on me to be a supervisor, plus quality controller and do all the packaging. And it just built up, built up. I just carried on, carried on with it, until the bubble burst. But for the first four months I didn't believe myself...I kept saying to the manager "I can't manage all this". And he goes, I can still remember his voice, he said "If you can't manage it find another job".'

The employee developed various physical problems including musculoskeletal pain associated with the packing. His symptoms included pains in elbows and wrists and his fingers went numb. He obtained a restriction note from the company doctor, stating that he could do certain jobs, but no lifting. His manager was unhappy with this situation and made things very difficult for him. Colleagues advised him to make a formal complaint but he chose not to. Eventually he consulted his GP about the physical symptoms but the GP also diagnosed depression:

'That's what my GP said to me the first day I went to him about it. And I just broke up, I just broke up in tears. I never used to believe that there was depression or anxiety, myself personally, until it just broke me down. It affected me like anything, and I've been off for a while now, and I get up out of bed every day, but when I tried going back to work I couldn't. I just could not face it.'

Figure 6.4 Example vignette from focus group report

Employees with responsibility for others (teachers, health care professionals, managers, mechanics and electricians) felt they were at particular risk with respect to the impact of anxiety and depression and their medication. People with responsibility for health and safety at work described how they failed to take action in connection with those responsibilities because they were too depressed and tired to respond.

A mechanic and an electrician described how they had to check and double check their work because they lacked confidence in their abilities. Health care professionals discussed how, as well as placing themselves at risk, they may have also posed a risk to their patients. Particular risks highlighted by health care workers included: impairment in clinical judgment, making clinical errors, unsafe behaviour during the handling of materials such as blood, and problems in the administration of drugs and needle injuries:

> *'I certainly noticed that when I'm not that great it can be very difficult to get blood out of people or get things into people without, you know, making a bit of a mess of it. The thing is if you're a psychiatrist like me and you're not doing these things every single day but you are doing them often enough to be proficient, what happens is you forget the routine of having stuff handy so you don't have the cotton wool. So you're standing there with a needle in somebody's arm and blood goes everywhere, all over you, all over the floor, over the patient and you're apologising and then you're realising "drug addict, oh goodness knows what they've got".'*
>
> (35-year-old female locum psychiatrist)

Non-compliance with medication was very common due to unpleasant side effects, lack of improvement in symptoms or because the medication initially made patients feel worse. A female university administrator explained how she experienced severe side effects and had to discontinue the treatment:

> *'I came off Prozac, even though the Doctor told me to stick with it, because I just felt so ill...my skin felt like it was alive and I was freaking out all the time. And tingles and palpitations and I just felt horrible...I'd rather be depressed or anxious. Tingles, it's weird, you can't explain...everything was completely exaggerated, and I was just freaking out completely, and I just couldn't eat. And not being able to eat always bothers me, so that made me more anxious. So I just thought "No, pack it in".'*

Some people took less than the prescribed amount of medication or stopped taking the medication altogether. People were also concerned about dependency and this caused them to cease medication prematurely. Employees were largely ill-prepared for their medication regimens and would have welcomed more information from doctors. As a 58-year-old male health and safety officer put it:

> *'I don't think there is enough explanation. I've seen three GPs and none of them actually explained what the tablets were that they were giving me. Just take them, they'll make you feel better after a few weeks.'*

It was remarked that leaflets accompanying medication often appeared to be serving as a disclaimer for the pharmaceutical company, rather that giving patients good insight into what to expect to happen. At work, individuals suffering anxiety and depression looked to managers and colleagues for support but often this was limited due a lack of understanding about their condition. A 58-year-old male health and safety manager who worked for a publishing company commented:

> *'I don't think senior management or personnel department actually know how to deal with it. They know what to do if somebody has a bad back from lifting, they stop them lifting...but when you're depressed and anxious and you can't do the job properly because there is no physical thing showing they don't know how to deal with it.'*

Drawing on the evidence collected, the research led to recommendations for the prevention and management of anxiety and depression in the workplace and possible improvements in health care provision for people suffering with these disorders. Primary recommendations were the need to raise general awareness of mental health issues, the importance of assessing workplace safety risks associated with symptoms and medication, and improving information for people taking medication in terms of what to expect.

Of particular interest in this case study was the benefit that individuals reported from participating in the focus groups, in terms of providing a source of mutual support and experience. On the other hand, the focus groups were emotionally demanding for the moderators, listening to traumatic experiences and with participants sometimes becoming upset. Having two moderators running the sessions helped with this, providing a degree of mutual support and debriefing. With the benefit of hindsight, this aspect of undertaking the research could have been more fully recognised at the outset, perhaps with improved provision for debriefing available.

6.4.3 Summary

This study provided extensive qualitative information concerning the impact on working life of mental health problems and their treatment. Although the effects of medication were difficult to separate from the condition under treatment, participants reported that both their work and safety were affected by one or the other. Opportunities were identified for improving the way in which those experiencing anxiety and depression are treated and supported. It is expected that the results will inform future guidelines for employers and employees, raising awareness of issues and setting out best practice for supporting employees with mental health problems and health and safety risk assessment.

6.5 CONCLUDING DISCUSSION

As illustrated by the examples presented in this chapter, focus groups are now an established methodology for health and safety ergonomics research. Elements of the scientific community, however, remain to be convinced that it is possible to learn anything useful from human studies lacking the rigorous experimental control to which they are accustomed. The method does not permit the proving of structured hypotheses and some academic colleagues are uncomfortable with anything less than this. Amongst clients, however, the approach is increasingly popular. This is in part due to the valuable insight focus groups provide into a research question, often sufficient for determining a way forward and allowing an issue to be addressed. In other circumstances, where there may be a need to persuade people as to the impact or severity of a situation, the personal accounts yielded by focus groups can be much more persuasive and compelling than quantitative data alone. Our anxiety and depression study is a good example of this.

One weakness in the focus group methodology is uncertainty concerning the extent to which participants and their views are representative of the study population under investigation. This problem can be overcome by using focus groups in combination with other methods, for example, survey work, observation or laboratory experimentation, and allowing triangulation. Another concern can be the subjectivity of focus group data analysis. This can be strengthened through the use of procedures for validating data interpretation, either by comparing the consistency of analyses undertaken independently, or through inviting representatives from the study group or other experts to comment on the conclusions reached.

The verbose nature of focus group data can present a challenge when reporting findings, with a balance to be struck between providing a comprehensive account of the data, whilst achieving this in a digestible form. The construction safety study discussed in this chapter demonstrated how data reduction into summary statements can be part of the analysis process. However, this does lose some of the richness provided by actual quotations. A combination of summarisation and illustrative quotes often works well.

As discussed earlier, a key lesson we have learnt from the use of focus groups is not to underestimate the time and resources needed to recruit participants. Attention to other practicalities is also important, such as making sure that the meeting venue is appropriate: quiet, private and easy for participants to find, for example. Another matter that can affect the smooth running of a group is guaranteeing sufficient time is available for relaxed discussion. We have encountered problems with this when gaining access to participants through a meeting already taking place for other reasons. In these situations, it is important when negotiating the session to emphasise the time required. Refreshments should be organised, although alcohol is best avoided. The possibility of participants drinking alcohol beforehand ought not be overlooked, especially where the gathering might be viewed by participants as an opportunity to socialise. We have had a boisterous session with a focus group conducted early in the afternoon, held with workers who already knew each other and with them having spent lunchtime in the bar. Finally, the importance of good quality audio-recording equipment should not be neglected. It can be frustrating, or indeed painful, to listen to tapes where it is difficult to hear the conversation or which have to be replayed at high volume. Attention to detail in focus group planning pays dividends down the line.

6.6 REFERENCES

American Geriatrics Society, 2001, Guideline for the prevention of falls in older persons. *Journal of the American Geriatrics Society*, **49**, pp. 664-672.

Brooke-Wavell, K., Jones, P.R.M. and Pye, D., 1995, Ultrasound and DEXA measurements of the calcaneus: influences of region of interest location. *Calcified Tissue International*, **57**, pp. 20-24.

Dunne, M., Hartley, L. and Fahey, M., 1986, Stress, anti-anxiety drugs and work performance. In *Trends in Ergonomics of Work*, 23rd annual conference of the Ergonomics Society of Australia and New Zealand, pp. 170-177.

Haslam, R.A., 2002, Targeting ergonomics interventions – learning from health promotion. *Applied Ergonomics*, **33** (3), pp. 241-249.

Haslam, R.A., Sloane, J., Hill, L.D., Brooke-Wavell, K. and Howarth, P., 2001, What do older people know about safety on stairs? *Ageing and Society*, **21**, pp. 759-776.

Hastings, S., Brown, S., Haslam, C. and Haslam, R.A., 2002, Effects of medication on the working population. In *Contemporary Ergonomics 2002,* McCabe, P.T. (ed.), (London: Taylor and Francis), pp. 415-420.

Health and Safety Commission (HSC), 2001, *Managing Health and Safety in Construction: Construction (Design and Management) Regulations 1994*, (Sudbury, Suffolk: HSE Books).

Hide, S., Hastings, S., Gyi, D., Haslam, R.A. and Gibb, A., 2001, Using focus group data to inform development of an accident study method for the construction industry. In *Contemporary Ergonomics 2001*, Hanson, M.A. (ed.), (London: Taylor and Francis), pp. 153-158.

Hill, L.D., Haslam, R.A., Brooke-Wavell, K., Howarth, P.A. and Sloane, J.E., 2000, *Avoiding slips trips and broken hips: How do older people use their stairs?*, (London: Department of Trade and Industry).

Morgan, D.L. and Krueger, R.A., 1998, *The Focus Group Kit*, (Thousand Oaks, CA: Sage Publications).

Nicholson, A.N., 1990, Medication and skilled work. In *Human factors and Hazardous Situations*, Proceedings of the Royal Society Discussion Meeting, 28-29 June, Broadbent D.E. (ed.), (Oxford: Clarendon Press), pp. 65-70.

Potter, W.Z., 1990, Psychotropic medications and work performance. *Journal of Occupational Medicine*, **32**, pp. 355-361.

Tilson, H.H., 1990, Medication monitoring in the workplace: toward improving our system of epidemiologic intelligence. *Journal of Occupational Medicine*, **32**, pp. 313-319.

Acknowledgements

Several colleagues in the Health & Safety Ergonomics Unit played a leading role in designing, executing and reporting the studies recounted in this chapter, in particular Denise Hill, Joanne Sloane, Sophie Hide, Sarah Hastings and Sue Brown. Alistair Gibb, Diane Gyi, Katherine Brooke-Wavell, Peter Howarth and Cheryl Haslam were co-directors of these projects. Funding from the Department of Trade and Industry and the Health & Safety Executive is acknowledged. The views presented in this chapter are those of the author and do not necessarily represent those of colleagues or sponsors.

CHAPTER 7

Running Focus Groups with Older Participants

Julia Barrett and Paul Herriotts

7.1 AN AGEING POPULATION

It is widely known that the world's population is ageing, with more developed countries leading the process. According to the United Nations Population Division (2001), globally the number of older persons (aged 60 or over) will more than triple by 2050. The increase in the number of 'oldest old' (aged 80 or over) is expected to be more dramatic, being more than five fold. The reason for this demographic trend is a rise in life expectancy and a decline in world fertility (United Nations Population Division, 2001).

The support of this ever-expanding older population has become of increasing concern, as has the need to give greater attention to the design of products and services to older consumers. Not without good commercial reason: it is these people who exercise the power of the so-called 'grey pound' (the Henley Centre and DesignAge, 1997).

In the UK, the trend for an ageing population, with low mortality rates and still lower fertility rates is evident. The Office for National Statistics (2001) predicts that these trends will continue and those aged 65 and over will exceed those under 16 for the first time in 2016. The balance of the population will change drastically. Proportionally, the biggest increase has been in the oldest old: the group most likely to have chronic illness and disability, and to need care.

7.2 AGEING AND DISABILITY

The prevalence of longstanding illness and disabilities increases with age. Older people are more likely to experience multiple impairment such as both failing sight and hearing, and chronic conditions such as osteoarthritis. In the General Household Survey (GHS) of 1998, 59 per cent of people aged 65-74 and 66 per cent of those aged 75 and over reported having a longstanding illness, compared with 33 per cent of people of all ages (Office for National Statistics, 2000). The GHS also reports that older people are likely to under-report chronic illness or disability.

Disability increases rapidly with age, with around half of all disabled people in the UK aged 65 and over. Only 4 per cent of those with a disability are 16-24 in age, compared to 23 per cent aged 65-74 and 26 per cent aged 75 and over (Department of Health, 1998).

7.3 OLDER PARTICIPANTS IN FOCUS GROUPS

Focus group practitioners such as Krueger (1994) and Morgan (1997, 1998) have provided guidelines on how to plan and run successful focus groups, based on their own and others' experiences. However, special considerations for older participants are lacking.

Barrett and Kirk (2000) conducted a series of focus groups with older and disabled older participants (aged 60 or over) and their carers (also mainly aged 60 or over). Focus groups were used to provide some insight into what older people and their carers wanted to make life at home easier, particularly in terms of information and advice and the reasons for this. A number of difficulties arose, most of which were associated with age-related declines in sensory, perceptual, cognitive and communication abilities, as described in Section 7.4 below. Practical steps that need to be taken to eliminate or reduce these difficulties are described in the guidelines section (Section 7.5). Adoption of these steps will enable this powerful method of collecting qualitative data to be used effectively with older people (Figure 7.1).

Figure 7.1 Running focus groups with older participants requires special considerations

7.4 EFFECTS OF AGEING ON VARIOUS ABILITIES AND THE IMPLICATIONS FOR FOCUS GROUPS

There are a number of sensory, perceptual and cognitive effects of ageing that can affect performance of tasks. Many studies (see Morrow and Leirer, 1997, for an overview) have shown age-related declines in a range of abilities including:

- visual and auditory perception;
- divided and focused attention;
- sustained attention;
- information processing in working memory;
- recall from semantic memory;
- speech processing;
- language production.

These declines in ability have implications for the running of focus groups with older participants.

Participant performance is more likely to reflect these changes under complex task conditions. Whilst it is useful to note areas of declining ability, it is important to remember that there are marked individual differences in the rate and magnitude of deterioration, and older people often find ways to compensate for these deficits (Howell, 1997). The changes described in this chapter relate to the normal ageing process and do not take into account impairments due to illness or injury, which could exacerbate any difficulties experienced when running focus groups with older participants.

Focus groups involve discussion. Thus, age-related declines in the ability to perceive, comprehend and remember speech and also to produce language have the most important implications for the performance of older participants. Decline in auditory perception, attention, speed of information processing and memory, all of which have an impact on spoken language communication, are described first in the sections below. Research related more specifically to speech processing and language production is then outlined. Finally, visual perception, processing of printed information and literacy, which have implications for using visual materials and media in focus groups, are covered.

7.4.1 Auditory and Speech Perception

Kline and Scialfa (1996, 1997) summarised age-related changes in the auditory system and auditory perception. In older people, physiological changes in the auditory system causes reduction in pure tone sensitivity, deficits in high frequency pitch discrimination, hearing loss at certain frequencies in the speech range and deficits in speech recognition. The loss of sensitivity to higher frequencies with advancing age (known as presbycusis) is greater for men than for women – this difference increases with age and is not related to differences in intensity of noise or duration of exposure.

Frequency discrimination, important in speech perception and many auditory tasks, declines with age. There is also a loss in the ability to localise sound with age. In studies using dichotic listening tasks, older people have difficulty repeating the information presented in one ear if competing information is presented to the other ear (e.g. Barr and Giambra, 1990). This suggests a divided attention deficit. Older people also need more time to switch attention from one sound to the other, causing them to miss information that younger people hear (Wickens *et al.*, 1987). These difficulties have important implications for the performance of older people in noisy environments or situations, such as focus groups, where more than one person may be speaking at the same time. Age-related deficits in speech perception, demonstrated by many researchers, are small when the speech occurs in a quiet environment, but increase markedly in noisy or reverberant environments (Tun and Wingfield, 1997).

7.4.2 Attention

Older people exhibit:

- a lowered ability to focus attention and inhibit or suppress irrelevant information, in both auditory and visual tasks (Morrow and Leirer, 1997; Tun and Wingfield, 1997);
- increased difficulty in dividing attention, especially for activities that require monitoring of complex or novel information, such as driving whilst reading road signs (Tun and Wingfield, 1997);
- increased difficulty in sustaining attention and performance over long periods of time (Vercruyssen, 1997).

In terms of running focus groups with older participants, these declines in attentional capacities could cause a number of difficulties for the moderator and later in analysis. The experience of Barrett and Kirk (2000) showed that older participants could have difficulty focusing attention on the particular topic currently under discussion. Some participants were unable to ignore irrelevant comments from others and their attention was diverted towards these remarks. In addition, some participants would recall a particular salient memory, perhaps having been influenced by a comment from someone else, and remain focused on this, continually referring to it throughout the discussion, even though it might have been irrelevant. Failure to attend to the current topic of conversation was the result. It was also found that participants experienced difficulty in sustaining attention beyond around one hour of continuous discussion or one and a half hours if a short mid-session refreshment break was included.

7.4.3 Cognitive Slowing

A popular theory is that age-related declines in ability are associated with a gradual slowing of processing speed (e.g. Salthouse, 1991). Researchers agree that the effects of ageing are greater on cognitive components of performance than on perceptual-motor components and a slowing of response speed with age is due more to central (cognitive) than peripheral factors (Vercruyssen, 1997). For older people speech processing is affected by speech rate. As speech rates increase, older adults are less able to discriminate, comprehend and recall spoken material than young adults (Kline and Scialfa, 1996; Tun and Wingfield, 1997).

Older adults need more time than younger adults to process information, to recall material from semantic memory, to retrieve the correct words in language production, and to plan what to say and how to say it.

Barrett and Kirk (2000) found that a participant would occasionally answer the moderator's question by giving a reply that was relevant to a previous question or even a previous topic, sometimes providing useful information relating to the earlier point of discussion. It appears that participants may have been continuing to think about a question and their response to it as the discussion proceeded, after other questions had been asked and even after the next topic had begun. Such events broke the flow of the discussion, and could cause other participants to divert their attention. The skills of the moderator in focusing and guiding the discussion were put to the test. Subsequent organisation of the resulting data so that it could then be described and interpreted also became a more complicated task.

7.4.4 Memory

Older people tend to report that their memories have become worse with age (Ryan, 1992). A large number of laboratory studies have confirmed this. Short-term memory tests that require people to retain small amounts of information for small periods show little or no age differences. However, working memory capacity (information stored and currently being used for an ongoing task) shows an age-related decline. Tests of working memory that require people to simultaneously manipulate or process the information show a performance deficit for older people, particularly when the processing demands are heavy (Howard and Howard, 1997). Working memory plays an important role in many activities, including visual and speech processing. As the complexity of both verbal and spatial working memory tasks increases, older adults show greater performance deficits than younger adults (Morrell and Echt, 1997). Thus, it is the processing component of working memory that is important in understanding age-related deficits.

The difficulty in inhibiting task-irrelevant information means that this information competes with more relevant information for working memory capacity. This makes older people more susceptible to disruption of material in working memory by interference when attention is diverted (Morrow and Leirer, 1997). Failure to inhibit irrelevant information results in: increased activation of material in working memory that is not directed towards the goal of the ongoing task; inefficient allocation of attentional resources; and comprehension and memory deficit for the target information (Tun and Wingfield, 1997).

Decay and interference of material in working memory due to diverted attention was observed in the focus groups run by Barrett and Kirk (2000) when a speaking participant was interrupted by other members of the group. When the moderator later returned to that person, they had forgotten what they had started to say, even when reminded of their first few words and the discussion leading up to their interrupted remark. In such cases, potentially useful information was lost.

Older adults also seem to keep irrelevant information activated for longer than younger adults (Hamm and Hasher, 1992). Barrett and Kirk (2000) found that some older participants were easily distracted, unable to ignore irrelevant comments and continued to remember irrelevant information, continually referring to it throughout the discussion.

Recall from semantic memory (our store of world knowledge) can be difficult for older people if processing time is limited. Whilst older people may have more so called 'tip-of-the-tongue' experiences (when a person is unable to retrieve some or all of the phonological aspects of a familiar word), if they are given time their accuracy of recall is as high as that for younger people (MacKay and Abrams, 1996; Howard and Howard, 1997).

Barrett and Kirk (2000) found that some participants had difficulties retrieving the names of local organisations that they had used for information or advice, but were able to describe the function and location in the town.

7.4.5 Speech Processing

Speech processing in older people is affected by a number of the changes in ability described above:

- declining ability to perceive auditory information;
- deficits in focused and divided attention;
- slowed information processing;
- working memory limitations.

The working memory limitations of older people have important implications for spoken language communication. Sentences with especially complex syntax cause problems. In addition, although speech recognition and memory in older listeners is helped by linguistic context, context cues that arrive after the poorly perceived speech have limited usefulness due to the burden of keeping the ambiguous speech in memory whilst awaiting the potentially clarifying information. Background noise and reverberant conditions can cause additional problems for older listeners. Care must be taken not to overload the working memory of older people in situations that require simultaneous maintenance and manipulation of spoken information (Tun and Wingfield, 1997). Barrett and Kirk (2000) found that some participants were unable to recall the contents of the discussion when asked at the end to state what they felt to be the most important point.

In the focus group study by Barrett and Kirk (2000), participants were provided with background information and an overview of the discussion topics prior to the discussion. In addition, the moderator introduced new topics to the participants before the relevant questions were asked. The questions were designed to be short, using simple language, and concerning one issue only. The wording of the questions was also tested on people similar to the target participants before use in the focus groups. Nevertheless, participants would occasionally sit in silence after a question was asked. It was unclear, initially at least, whether the participants had not heard the question, did not understand it, needed time to give their response, or did not wish to answer. Calling on an individual could cause them to give a response that was irrelevant to the question just asked or was relevant to a previous question. In such cases, it helped the participants to understand what was wanted if the moderator provided cues by speaking of his or her own experiences before repeating the question.

In relation to many other cognitive abilities, speech comprehension and memory are well preserved in old age due to the ability of older people to compensate for deficits in lower level sensory or memory capability. For example, in addition to using context cues, older people benefit from the rich linguistic structure of natural speech; especially in difficult listening conditions (Wingfield *et al.*, 1995). The prosodic patterns of speech also aid in comprehension and recall. Older people also compensate for deficits in sensory abilities ('bottom-up' input) by relying more on the predictability of communication and use their knowledge of past experience to provide cues for comprehension and memory ('top-down' processing) (Tun and Wingfield, 1997). However, this processing style means that older people often inappropriately apply top-down processing and personalised knowledge as routine (Willis, 1996). This could account for some of the irrelevant responses given by older participants to moderator's questions in the focus groups run by Barrett and Kirk (2000).

7.4.6 Language Production

In terms of language production, older people exhibit deficits in two fundamental areas: word retrieval and planning what to say and how to say it. Older adults often report problems with retrieving familiar words and studies have shown that the spontaneous speech of older people contains more pronouns and ambiguous references and is slower due to both word lengthening and longer and more frequent pausing. From about age 37 the frequency of tip-of-the-tongue experiences increases with age, both in naturally occurring speech and in laboratory settings (Burke *et al.*, 1991). Occurrences of disfluencies in speech (hesitations, false starts and word repetitions) also increase with age, indicating deficits in language planning (MacKay and Abrams, 1996).

In a focus group situation, these factors will contribute to a slowing down of the pace and disruptions to the flow of the discussion.

7.4.7 Visual Perception

Kline and Scialfa (1996, 1997) summarised the effects of ageing on the visual system and visual perception. Owing to physiological changes in the eye, ageing is associated with a decline in static visual acuity, particularly in low light levels, and a decline in contrast sensitivity. Loss of visual acuity and contrast sensitivity has implications for the design and use of printed materials or visual media in focus groups with older participants. Thus, for older people, increased illumination and contrast enhance legibility. However, whilst older people can benefit from increased light levels, they have increased susceptibility to glare and a longer recovery period after exposure.

7.4.8 Processing of Printed Information

Age-related declines in visual perception, spatial abilities and working memory have implications for the processing of written or printed information. It has been suggested that there is an increased reliance on memory to compensate for the decreased quantity and quality of the information entering the older visual system (Morrell and Echt, 1997). Age-related deficits in working memory, described above, are well documented.

As with other abilities, the performance of older adults in processing printed information is adversely affected by high working memory or information processing demands. Studies suggest that older people have little difficulty with typographically simple text or text that is familiar to them, but experience considerable problems with typographically complex text that deals with unfamiliar material (see Hartley, 1994, for a summary).

These findings have implications for the preparation of printed materials in focus groups with older participants.

7.4.9 Literacy

In a 1994 small-scale survey of the basic skills of different age groups in England and Wales by Gallup (the Basic Skills Agency, 1996) the oldest age group (72-74) performed worse on average than the younger age groups. However, in a self-report part of the study, this age group had the fewest reported difficulties. The report suggests that it is likely that the poor performance of the oldest age group in the literacy tests is due to the effects of ageing, although there is no way of knowing whether the people were more competent when they were younger.

A 1996 Office for National Statistics' small-scale survey of adult literacy in Britain (Carey *et al.*, 1997), found that the highest incidence of poor literacy was amongst the oldest group examined, those aged 55-65.

These findings have implications for the use of written materials in focus groups with older participants.

7.5 GUIDELINES

Focus groups with older participants test the design and organisation of the questions and the skills of the moderator. With older participants, one should plan for shorter focus group sessions, of a slower pace, with more disruptions to the flow of the discussion and, hence, more moderator involvement and effort.

7.5.1 Question Design

The questions must be simple, short, and use words that older participants understand. The wording of the questions can be tested out on older people similar to the target participants prior to an actual focus group. The examples below, taken from the focus group project run by Barrett and Kirk (2000), show questions before testing with target participants and the final questions used in the focus group sessions.

Draft question:
'Should the publication be providing brief advice or telling people how and where to get the advice?'
Revised question:
'What information and advice would you like to see in the new publication?'

Draft question:
'Would you like to see products designed specifically for problems experienced by elderly people or products that are considered to be handy for everyone or both?'
Revised question:
'Would you like the publication to provide information on useful products and changes to the home?
What information would you like?'

As always, careful preparation is vital: this involves not only the design of the questions, but also their organisation and introduction to help focus the discussion. Focusing is aided by starting the discussion with a general introduction to the topic, followed by general questions, with a progression to more specific ones. It is particularly useful with older participants to provide background information about the purpose of the study and to establish the context of the questions. This can be achieved in the invitation letter as well as the moderator's introduction.

The early stages of a focus group session are especially important with older participants (the welcome, the overview of the topic, the ground rules and the first question). Recognition, comprehension and memory of spoken information in older people benefit from prior context cues (Tun and Wingfield, 1997). Thus, establishing a context for the questions before the discussion begins, and also before sub-groups of questions, will help older participants to process the subsequent questions.

Barrett and Kirk (2000) found that although their questions were divided into groups of two to six questions and the topic of each group was described beforehand, on occasions the participants still had problems understanding questions. More frequent context cues are recommended, possibly before every question asked. An example is shown below.

> 'The first set of questions is concerned with information and advice. I'll just speak from my own experience: when I was really unwell, I didn't know where to go for information, but nor did my wife, because she'd got no experience.
>
> Have you ever contacted an organisation or local centre for information or advice? If yes:
>
> - What was the information or advice that you were seeking?
> - Who did you contact?
> - Did you make contact by letter, telephone or visit?
> - Did you have any problems getting the information and advice you wanted? If so, what were the problems?'

7.5.2 Participant Recruitment

For people of any age the motivation to attend may be increased by the provision of an incentive such as payment or an appropriate gift. However, such incentives will have little effect when circumstances beyond an individual's control prevent attendance. This is more likely to occur with older participants, due to illness, hospital appointments or the demands of looking after a disabled partner. This needs to be considered when recruiting older participants. It is always necessary to over-recruit to allow for people not showing up on the day and Morgan (1997) recommends over-recruiting by 20 per cent. The experience of Barrett and Kirk (2000) suggests that it is wise to increase the extent of over-recruitment when planning focus groups with older participants. A figure of 25 per cent is recommended.

Figure 7.2 The use of writing materials helps ensure that participants' thoughts are captured

7.5.3 Moderator Skills

Older participants present challenges to the moderator. The moderator needs to be especially adept and practised in communicating clearly, listening carefully and sensitively, guiding the conversation, encouraging responses, controlling the participants, making participants feel that their responses are valued, and keeping the discussion focused on the aim of the study. Although younger moderators were not used in Barrett and Kirk's study (2000), those of a similar age to the participants were found to be effective in putting them at ease. By referring to their own life experiences, older moderators were able to encourage participants to speak about personal experiences and to provide cues to questions that participants were unable to answer.

The moderator needs to speak clearly and slowly. Older participants must be allowed time to think about a question and to respond to it. The moderator should try to ensure that only one person speaks at a time to allow for difficulties that older people experience in dividing attention between more than one speaker. If more than one person does speak, then the moderator should ask each to repeat what they have said in turn. Interruptions should be kept to a minimum, with participants allowed to complete a comment, rather than allowing interruptions and returning to the interrupted participant later on. If the participant later forgets what they had started to say, the researcher is left with a tantalising few words that may have gone on to provide useful information to the study. Providing participants with writing materials is recommended. Participants should be encouraged to write down any thoughts that they have whilst someone else is talking (Figure 7.2).

7.5.4 Printed Materials

Printed materials are often used in focus groups. When preparing these for older participants, general guidelines on designing text to improve legibility and reduce working memory demands for older readers should be considered (e.g. Hartley, 1994; Morrell and Echt, 1997). Recommendations include:

- using a large text size of 12-14 point;
- using a sans serif font style;
- having a clear text structure (with spacing, headings, numbering, etc.);
- having a clear layout;
- using directive cues (e.g. bold type, italics);
- avoiding use of all capitals in body text;
- using simple language.

Printed materials should be tested out on people similar to the target participants before use in a focus group. However, careful consideration should be given to whether reading materials are absolutely necessary: those with poor literacy skills may feel extremely uncomfortable and resentful if made to carry out a task involving reading in a group situation. Of course, focus groups consisting purely of discussion will not discriminate against those with poor literacy.

7.5.5 Environment

A comfortable environment is important (Figure 7.3). The lighting should be adequate for the participants to read easily if visual media are going to be used (e.g. flip-chart, handouts, questionnaires, display screens). However, care should be taken to ensure that glare is avoided. Again the environment can be tried out with older people, prior to an actual discussion.

A quiet and echo-free environment is essential. Reverberation in a room can be reduced by the presence of carpets, curtains and plants. Other considerations are an easily accessible location that does not require climbing flights of stairs, comfortable seating, and a warm room temperature. If participants have special access needs, the focus group should be held in a suitable location to accommodate them, e.g. with entrances and toilets that enable wheelchair access. If a number of hearing aid users are to participate, provision of an audio induction loop system should be considered to help them hear more clearly.

7.5.6 Before the Session Begins

Participants should be met at the venue entrance and greeted by the moderator and assistant moderator. They should be made aware of the location of the toilets. It is important to engage the participants in conversation prior to moving into the meeting room or sitting down for the discussion. This can be used to identify individuals with hearing difficulties, who should then be seated nearest to the moderator.

Participants should be asked what they prefer to be called, and these names written in large letters on name cards or labels so that the moderator and participants are relieved of the burden of remembering everyone's name. The moderator should also be identified in this way.

Figure 7.3 A comfortable environment is important

7.5.7 Duration of Discussion

The discussion should be kept short. A maximum of one and a half hours is recommended, if a short mid-session refreshment break is offered. If the discussion is continuous, without a break, then a maximum duration of one hour is recommended. It is a good idea to keep participants continually updated on the time remaining.

Because the discussion will be short and run at a slow pace, the researcher should expect to cover fewer topics in a single focus group session run with older adults than in one run with younger adults. It may be necessary to run more sessions to gather all the information required.

7.5.8 Task Complexity

With any age-related decline in ability the following applies: the more complex the task the more performance will suffer. Therefore, when using older people in focus group discussions the recommendation is – keep it simple.

7.6 CONCLUSIONS

Focus groups with older participants can be run successfully using the guidelines given in this chapter and summarised in Figure 7.4. Useful qualitative information can be obtained thus making focus groups a vital tool in the design process. In a focus group study conducted by Herriotts *et al.* (2000), it was found that appropriately conducted focus groups were highly successful in generating vehicle design issues of importance to older car users. Moreover, as older people are a sensitive test of a design, using their input in the design process will result in products that are suitable for young and old alike. Use of older people as a design resource in well-run focus groups is therefore strongly recommended.

Careful preparation is vital.

Questions must be simple, short and easy to understand.

Establish the context of the questions:

- introduce participants to the topics that will be discussed in the invitation letter and in the moderator's introduction to the discussion;
- divide questions into sets of related questions; during the discussion introduce new topics prior to the relevant set of questions.

Recruitment: incentives help, but over-recruit by 25 per cent just in case.

Moderator: preferably older, with good moderating skills.

Environment:

- easily accessible;
- well lit;
- quiet and echo free;
- comfortable temperature;
- comfortable furniture;
- provide an audio induction loop system if necessary.

Provide writing materials for participants.

Only use printed material if absolutely necessary.

Use large size text in a san-serif font style, with a clear structure and layout.

Duration: one hour maximum, or one and a half hours (with a short break), run at a slow pace.

Test everything with older people prior to the actual focus group.

KEEP IT SIMPLE!

Figure 7.4 Summary of guidelines for running focus groups with older participants

7.7 REFERENCES

Barr, R.A. and Giambra, L.M., 1990, Age-related decrement in auditory selective attention. *Psychology and Aging*, **5**, pp. 597-599.

Barrett, J. and Kirk, S., 2000, Running focus groups with elderly and disabled elderly participants. *Applied Ergonomics*, **31**, pp. 621-629.

Basic Skills Agency, 1996, *Older and younger: The basic skills of different age groups*, (London: The Basic Skills Agency).

Burke, D.M., MacKay, D.G., Worthley, J.S. and Wade, E., 1991, On the tip of the tongue: What causes word finding failures in young and older adults? *Journal of Memory and Language*, **30**, pp. 542-579.

Carey, S., Low, S. and Hansboro, J., 1997, *Adult literacy in Britain*. Office for National Statistics, (London: The Stationery Office).

Department of Health, 1998, *Health and Personal Social Services Statistics for England, 1998 edition*, (London: The Stationery Office).

Hamm, V.P. and Hasher, L., 1992, Age and the availability of inference. *Psychology and Aging*, **7**, pp. 56-64.

Hartley, J., 1994, *Designing Instructional Text* (3rd edition), (London: Kogan Page Ltd).

Henley Centre and DesignAge, 1997, *The older marketplace: a goldmine but still a grey area*. Presentation by the Henley Centre and DesignAge (Royal College of Art) at The Royal College of Art, London, March 1997.

Herriotts, P., Crawford, J. and Nayak, L., 2000, Engineered for whom? The transgenerational approach to car design. In *Proceedings of the 6th Annual* Research *Symposium*, School of Manufacturing and Mechanical Engineering, University of Birmingham, May 2000.

Howard, Jr., J.H. and Howard, D.V., 1997, Learning and Memory. In *Handbook of Human Factors and the Older Adult*, Fisk, A.D. and Rogers, W.A. (eds), (San Diego: Academic Press), pp. 7-26.

Howell, W.C., 1997, Forward, perspectives and prospectives. In *Handbook of Human Factors and the Older Adult*, Fisk, A. and Rogers, W.A. (eds), (San Diego: Academic Press), pp. 1-6.

Kline, D.W. and Scialfa, C.T., 1996, Visual and auditory aging. In *Handbook of the Psychology of Aging* (4th edition), Birren, J.E. and Schaie, K.W. (eds), (San Diego: Academic Press), pp. 181-203.

Kline, D.W. and Scialfa, C.T., 1997, Sensory and perceptual functioning: basic research and human factors implications. In *Handbook of Human Factors and the Older Adult*, Fisk, A.D. and Rogers, W.A. (eds), (San Diego: Academic Press), pp. 27-54.

Krueger, R.A., 1994, *Focus Groups a Practical Guide to Applied Research*, (Thousand Oaks: Sage Publications).

MacKay, D.G. and Abrams, L. 1996, Language, memory, and aging: distributed deficits and the structure of new-versus-old connections. In *Handbook of the Psychology of Aging* (4th edition), Birren, J.E. and Schaie, K.W. (eds), (San Diego: Academic Press), pp. 251-265.

Morgan, D.L., 1997, *Focus Groups as Qualitative Research* (2nd edition), Qualitative Research Methods Series, Vol. 16. (Thousand Oaks: Sage Publications).

Morgan, D.L., 1998, *The Focus Group Guidebook*. Focus Group Kit 1, (Thousand Oaks: Sage Publications).

Morrell, R.W. and Echt, K.V., 1997, Designing written instructions for older adults: learning to use computers. In *Handbook of Human Factors and the Older Adult*, Fisk, A.D. and Rogers, W.A. (eds), (San Diego: Academic Press), pp. 335-361.

Morrow, D. and Leirer, V., 1997, Aging, pilot performance and expertise. In *Handbook of Human Factors and the Older Adult*, Fisk, A.D. and Rogers, W.A. (eds), (San Diego: Academic Press), pp. 199-230.

Office for National Statistics, Social Survey Division, 2000, *Living in Britain: Results from the 1998 General Household Survey*. (London: The Stationery Office).

Office for National Statistics, 2001, *Social Trends 31, 2001 edition*, (London: The Stationery Office).

Ryan, E.B., 1992, Beliefs about memory changes across the adult life span. *Journal of Gerontology: Psychological Sciences*, **47**, pp. 41-61.

Salthouse, T.A., 1991, *Theoretical perspectives on cognitive aging*, (Hillsdale, NJ: Erlbaum).

Tun, P.A. and Wingfield, A., 1997, Language and communication: fundamentals of speech communication and language processing in old age. In *Handbook of Human Factors and the Older Adult*, Fisk, A.D. and Rogers, W.A. (eds), (San Diego: Academic Press), pp. 125-149.

United Nations Population Division, 2001, *World Population Prospects: The 2000 Revision*, www.undp.org/popin/#trends (accessed July 2001), (New York: United Nations).

Vercruyssen, M., 1997, Movement control and speed of behaviour. In *Handbook of Human Factors and the Older Adult,* Fisk, A.D. and Rogers, W.A. (eds), (San Diego: Academic Press), pp. 55-86.

Wickens, C.D., Braune, R. and Stokes, A., 1987, Age differences in the speed and capacity of information processing: 1. A dual-task approach. *Psychology and Aging*, **2**, pp. 70-78.

Willis, S.L., 1996, Everyday Problem Solving. In *Handbook of the psychology of aging* (4th edition), Birren, J. E. and Schaie, K. W. (eds), (San Diego: Academic Press), pp. 266-307.

Wingfield, A., Tun, P.A. and Rosen, M., 1995, Age differences in veridical and reconstructive recall of syntactically and randomly segmented speech. *Journal of Gerontology: Psychological Sciences*, **50B**, pp. 257-266.

CHAPTER 8

Bringing Real World Context into the Focus Group Setting

Peter Coughlan and Aaron Sklar

8.1 INTRODUCTION

This chapter examines the ways that IDEO, a design consulting firm, has adapted the standard focus group methodology in order to better inform its own process. These adaptations introduce elements of the 'real world' into the focus group setting, and extend the focus group setting back into the real world. We provide examples from actual project work with our clients to illustrate how our adaptations have led to a more effective design research methodology.

Focus groups have long been an important tool in the market researcher's toolkit. Most of what we have come to think of as the 'standard' focus group methodology has been developed through this market-focused application. As design practitioners, however, we have found the standard focus group methodology to fall short as a useful design tool, for two important reasons.

First, the standard focus group relies heavily on what people say they do or think they do, rather than what they actually do. In our experience, we have found verbal accounts of behaviour to be less accurate than observation of actual behaviour in real world settings. Humans are great confabulators – when put on the spot, we can usually come up with a verbal explanation of a particular behaviour or feeling ('Oh, I do/feel *X* because of *Y*'). When we scratch below the surface explanation, we can often find some other variable at play, one that we are not able to articulate because we are not conscious of it.

Second, focus groups cut people off from the contexts where they live their lives. People use their environments and tools to prompt them as to the next step in a sequence of steps; places prompt memories of past experiences. Sequestered in a barren room with a handful of strangers, people are stripped of some important resources that would ordinarily help structure their behaviours. Without these resources, we must rely solely on their accounts of their behaviours.

8.2 SOME BACKGROUND ABOUT IDEO

IDEO is a design consulting firm that helps its clients to create new experiences through the creation of products, processes, environments and communications. Fundamental to our process is a deep understanding of how our clients' customers or other stakeholders (such as employees, collaborators and users) behave in the real world. Most business people only know their stakeholders as members of particular market segments, with a specific demographic or psychographic profile; few spend time seeing these stakeholders as whole people, watching them experience life (or their company's products or processes) in everyday settings.

At IDEO, designers and design researchers inform the design process by looking at real world behaviour – that is, by understanding and observing behaviour in the everyday conditions in which design solutions will be adopted and adapted by various users. At the onset of a project, research helps the design team to *discover* unique and innovative opportunities and concepts. Through observation of people in everyday contexts, engaged in everyday activities, we seek to reveal people's unmet needs, or the ways in which existing designs fail to support their behaviours.

During design development, research is primarily employed to *test* prototypes of potential concepts by observing how users react to and employ the prototypes during tasks we ask them to perform. The findings from this prototype testing then inform subsequent refinement of concepts.

During the discovery phase of our work, market-research-style focus groups are not a mainstay of our design research process. However, during prototype testing, we have developed numerous variations on the focus group methodology to help inform our design process in ways consistent with our earlier methodologies.

The following pages describe our experiences with the use of focus groups, and provide suggestions about how to augment the more traditional focus group methods to better support the design process. Underlying all of our changes to standard focus group methodology is a desire to re-introduce *context* to the somewhat decontextualised setting of the traditional focus group – or, in other words, to augment the focus group setting with some of the trappings of the real world, and people's real-world behaviour.

8.3 FIRST AND FOREMOST, USE FOCUS GROUPS FOR THE RIGHT REASONS

Perhaps the primary reason that focus groups have gained so much popularity as a research tool in product development is that they provide a means for members of a development team to get relatively quick input or feedback about a potential design direction. In a period of a couple hours, the team can get a qualitative understanding of multiple potential customers' personalities, as well as their reactions to a series of design concepts or directions. In order to bring eight to ten individuals together for a discussion, however, we must often ask research participants to check their individuality at the door. In the space of two hours, there is little opportunity to explore some of the very things that designers find so interesting and inspiring – people's passions, quirks, pet peeves, rituals, or the behaviours they have developed to cope with a poorly-designed product, service, or environment.

Although these (seemingly mundane) behaviours and beliefs often provide the inspiration for innovation, they rarely surface in a traditional focus group setting. For this reason, IDEO believes that focus groups are better used in some parts of the design process than in others. In general, we counsel our clients to use focus groups not for *discovery*, but rather to *test* ideas that have been created after new concepts have been discovered through other means. Discovery is triggered by looking at how people interact with one another and with designed objects in real-life settings; in how they engage with existing products, services, environments, communications, and systems; and in how they cope with change on a day-to-day basis. This kind of information is difficult, if not impossible, to capture in a focus group setting, where research participants have no access to the very artefacts that trigger and support their day-to-day behaviours.

One client we worked with had conducted focus groups at the onset of a design development program in order to help them with the design of a new website that offered web-based software services. The moderator for these sessions asked some potential software customers to describe 'the perfect online software service'. Participants spent the better part of two hours talking vaguely about good service, general brand values, and what services they thought they might need. At the end of the two hours, the client's general direction had been validated – confirming what they had already suspected before the start of the groups. Yet no specific direction had been identified as to how the service(s) could or should be designed or integrated into existing needs and behaviours. Nor was the client any closer to understanding how to create an interface for their users, or how to manifest their brand in that interface.

For answers to these questions, our design team sought out people whom we felt might need the services our client had to offer. We went into their offices and were given guided tours of their physical and virtual desktops; we visited our subjects' favourite websites, and shadowed them as they worked. We looked for recurring patterns and problems in work processes – patterns and problems that a new service or set of services could seek to build upon or address. With a better understanding of the end-users' daily lives, we were then able to create concepts informed by the types of software and services already being used, the environments where work was being done, and the work practices of the potential users of the new services being designed.

By conducting a small number of contextual observations, we collected fewer overall perspectives than we would have in a focus group setting. However, we came to know each person we observed in much greater depth than a focus group could provide us. Designers like to know minute details about their target users, so that they feel they are designing for a real person and not a profile of a target market group. In the case of the project described above, the design team learned how users organised and decorated their desktops, how they filed physical and electronic documents, how they navigated from program to program, etc. By understanding a user or group of users at this level of detail, the designers were able to generate insights that responded to a real set of users' unarticulated needs and aspirations. These needs and aspirations provide a valuable source of innovation because, before our process of discovery, they had *not* been articulated before.

In the focus group setting, participants do not have the contextual prompts available to us in our day-to-day existence, and must therefore recall their behaviour from memory. In the course of our field research, we have frequently observed that people do not do what they *say* they do; nor do they do what they *think* they do. There are many reasons why people's opinions do not match their behaviour: they may be on automatic pilot whilst performing a task, they may be telling you what they think you want to hear, or their memory may not be perfectly accurate.

Whatever the reason, it is clear that observation of actual behaviour is much more effective than recitation of recalled behaviours. So much can be learned from seeing body language and other behaviours that people are either unaware of or have forgotten were interesting. For example, whilst developing a treatment program for children with diabetes, we visited families who were already dealing with the condition. We watched a mother administer an injection to her young son and saw the routines they had developed together. The way that the mother held her son's hand whilst his arm rested on her leg was so intricate that it could not possibly be described verbally – it had to be witnessed as behaviour. First-hand observation of the relationship between mother and son reminded the design team to consider the emotional and psychological implications of the health condition in addition to the physical implications. With these insights, we created an aesthetic solution in keeping with the emotional tone of the parent/child interactions we observed. Needless to say, we would probably not have observed this behaviour during a focus group!

8.4 ASSIGN HOMEWORK BEFORE THE SESSION

One way to compensate for the relative lack of context found in the typical focus group setting is to ask participants to bring some context with them. By asking participants to bring something of themselves, they reveal at least some of their preferences, possessions, or everyday behaviours. For example, in a focus group we facilitated for a toy manufacturer, we asked children and their parents to bring in their favourite toys, and we began the session by having everyone talk about why they had chosen to bring in that particular toy (see Figure 8.1). As participants showed their toys to others in the group, we were able to learn the role of toys in their lives and what they valued about toys (as reflected in their choice of toys to bring to the focus group meeting). Similarly, in a project about handheld devices, we asked participants to bring in an object that they thought was pleasant to hold in order to learn what values and variables would apply to their assessment of the handheld prototypes we would be showing them later in the evening. The objects that participants chose to bring in also taught us a lot about their aesthetic preferences as well.

Figure 8.1 Asking parents to bring in their most and least favourite toy triggered memories and stimulated discussions amongst participants, and gave the design team specific and tangible insights into customers' value judgments about playthings for their children

In addition to asking participants to bring objects with them to a focus group session, we have found that asking them to perform an activity in advance can be an effective technique to prepare for a focus group. For an office re-design project, we asked participants to take pictures of key locations, objects, and interactions in their workplace. These images served as reminders or prompts of what was important to participants once they were away from the workplace setting, and helped focus group attendees to understand and respond to others' observations and insights as well.

Another useful preparatory activity is to encourage participants to keep a journal. In a project on mini-van interiors, we asked participants to keep a 'driving log' for a three-day period before the focus group meeting. They were asked to make note of the trips they went on, who rode in the car with them, their destination, how well the vehicle served their needs (for carrying passengers and people), and their emotional state during each trip. They then brought these journals to the focus group session, where the notes and records served to remind them of their experiences.

In each of these examples, a relatively open-ended (but carefully considered) homework assignment gave our participants the opportunity to study themselves, to reflect on their everyday behaviours, and to formulate a point of view about a design issue by focusing on it as they went about their everyday lives. Once in the focus group setting, participants could use their homework to jog their memories or to more effectively show others what they were referring to. As a side benefit, the quality of participants' homework assignments usually reveals who is most interested in, or even passionate about, the design topic; and, thus, who might have the most insightful contributions to make to the research. If more participants have shown up than you need (which is often the case since vendors routinely over-recruit to make sure they reach their participant quota), you can use the homework to evaluate the quality of insight from potential focus group participants.

8.5 DO NOT JUST ASK PARTICIPANTS TO *TALK* – GIVE THEM SOMETHING TO EXPERIENCE

Focus group moderators are really good at getting people to talk. In fact, the moderators with the best reputations in the industry are likened to talk show hosts because of their ability to fill a two-hour focus group session with non-stop conversation. We have found that whilst most people expect participants to talk during focus groups, specifically *non-verbal* activities can help participants 'articulate' an opinion about a design issue much more clearly than any amount of talking. This sometimes creates conflict with clients, who believe they are paying to hear people talk, and therefore object to any time participants spend *not* talking during a focus group session! For example, if you ask people to describe the right shape for the grip on a toothbrush, they will tell you it should be 'thick' or 'thin' or 'tapered' or 'straight.' Provide them with a range of prototypes of different grip sizes and shapes, blindfold them, and ask them to select the one that feels best to them, and they will give you much more refined and useful feedback.

Offering participants something tangible to experience during a focus group session leads to a deeper understanding of concepts and more spontaneous and therefore 'honest' reactions. Sketches and other visual stimuli are clearly a step in the right direction, but we have found physical prototypes to be the most powerful stimuli. It is easier to understand a product, environment, process, or message that can be held, walked through, or otherwise experienced through bodily interaction. It is also more satisfying for the participants (after all, they get to 'try' something) and more reliable for the researcher (who witnesses a physical response to something tangible, rather than a verbal interpretation of a concept described in words or pictures). In general, the more engaging the prototype, the more useful the information participants will be able to provide.

In situations where concepts have not been developed enough to create tangible prototypes, researchers can offer other tools to help the participant to communicate through non-verbal means. One such method we have used in the past involves the creation of collages. With this method, participants organise a collection of visual images and words (typically provided for them by the researcher) into some visual form that conveys their interpretation of a given design concept or situation. This technique is useful when participants are being asked to express their understanding or point of view of an abstract question. For example, in an effort to better understand the role of music in people's lives, we asked participants to create a collage around that topic. By discussing which images they selected and how images were arranged in the collages, we were able to solicit very rich and concrete conversations about an abstract topic. After participating in such an activity, participants often thank us for helping them to articulate their point of view about an issue they had never before verbalised.

8.6 LET PARTICIPANTS TAKE THINGS 'HOME' WITH THEM

Just as it is constructive to ask participants to bring some of the real world with them to the focus group setting, it can be beneficial to have them bring a design concept introduced in a focus group back into the real world – for example, their home, workplace or vehicle. In this way, participants can see for themselves how the concept fits into their everyday environment, with an eye towards collecting insights about how the design could be improved for their unique needs.

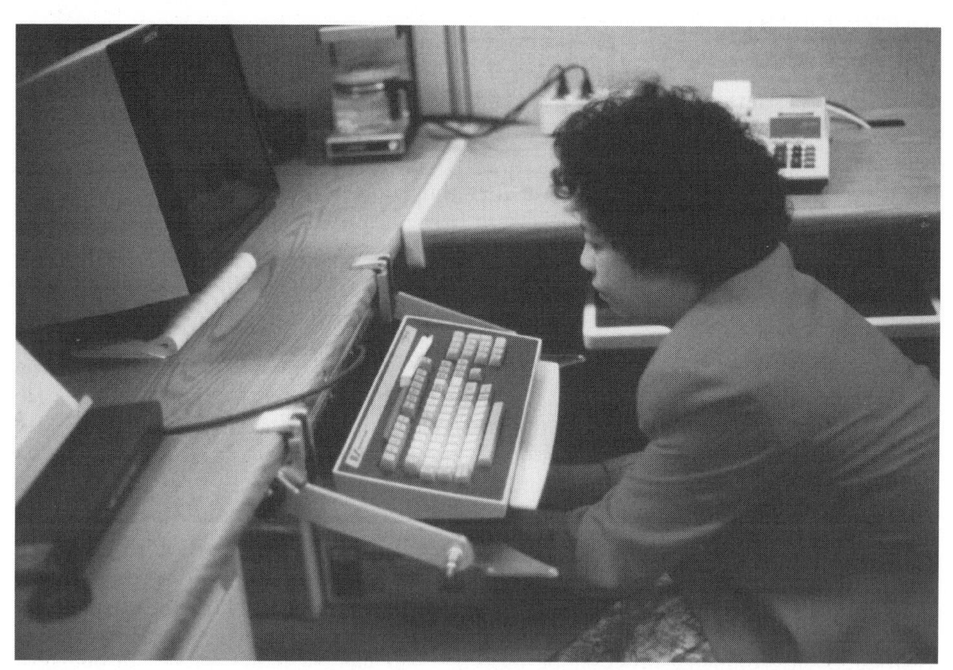

Figure 8.2 When research participants took prototypes of a keyboard support into their work environments, we learned that what they *said* they desired differed from what the environment would support. With this insight, we were able to quickly alter our design focus

In one past project, our client had discovered that their customers wanted a keyboard support mechanism that they could install and move themselves, without the help of a facilities person. We developed prototypes of such a self-installable and adjustable concept, and sent them out with users to be tested in their actual work environments (see Figure 8.2). Once users began using the product, we quickly realised that being able to move the keyboard support once it had been installed was a feature that users *thought* would be beneficial, whilst in reality most people had only one location to put their keyboard support. With this feedback, we quickly abandoned our efforts to create a feature that was quite expensive and in the end not very useful for customers.

One of our clients conducted focus groups around a new type of packaging for a milk-shake product. During the focus group, they asked participants to prepare the milk shakes (at the conference table, using plastic spoons provided by the moderator). When the moderator asked who in the family might like the new product, and when and how often they might consume the product, participants told the moderator what they thought would happen in their homes. Had they been instructed to take home samples of the product with a diary to record how they prepared the product, how and when they served it, and who ate which flavours, they would have learnt not what people said they *might* do, but what they *actually* did. In this way, they might have learnt that the containers did not fit well in the average freezer (suggesting the need for a different-size container); or that people preferred to drink from a glass rather than the paper cup that was provided (suggesting the potential for lower-cost packaging material). Such insights can only be gained by examining a concept in use in an everyday, real-world setting. Unfortunately, owing to the confidential nature of the project, we were unable to do such in-home evaluation at that stage in the design process.

When you cannot send a concept home with a participant (usually for reasons of confidentiality), a substitute activity involves asking participants to remember a concept after they have left the focus group room, and to test the memory of that concept within the context of their everyday lives. In one design project for an automotive manufacturer, we asked participants to make a brief audio note (we provided them with small handheld tape recorders) when they recalled the concept we had introduced them to during the focus group. Needless to say, the concept – a full-scale automobile – was too big to bring home! In a subsequent meeting, participants shared notes they had recorded – other vehicles on the road that reminded them of the concept they had seen; or cargo they were hauling around that weekend that would or would not have fitted into the concept vehicle. In addition, they recorded emotions that were triggered from memories of the concept vehicle, as well as the events that triggered these emotions. Although participants could not live with the actual prototype, they were able to introduce it into their lives and teach us a lot about how the concept fitted into those lives. This information helped the design team to refine the concept based on real-world feedback.

8.7 CONCLUSION

The context-enhancing methods recommended in this chapter are all geared towards transforming traditional focus groups into an activity that generates reliable, meaningful data for design. These methods create several challenges for the focus group moderator, because they require someone who can get participants to focus on actual, rather than desired or remembered, behaviour. This involves training the moderator to prompt for specific past instances of a behaviour, rather than a generic description of a behaviour. Other challenges to creating context-rich focus groups include the added cost and time needed to create and administer pre- or post-focus group homework tasks. The creation of 'props' (prototypes, everyday products, or visuals) to enrich the focus group experience also requires time and expense. In spite of the cost of these additional efforts, however, we have found them to be worth the expense because they provide the design team with a richer and more convincing view of the people for whom they are designing.

CHAPTER 9

Focused Groups: Scenario-based Discussions

Lee Cooper and Chris Baber

9.1 INTRODUCTION

Focus groups are a favourite tool for gathering users' comments on future product or service concepts. However, their suitability for this purpose is questionable. This chapter details a scenario-based focus group procedure that addresses a number of the problems that are associated currently with focus groups. It goes on to propose that 'focused groups', in which the discussion of concepts is concentrated on the development and sharing of scenarios, offers many benefits beyond the aggregation of individuals' opinions.

'In some ways the use of future fantasy has become an industrial design ethic in its own right' (McDermott, 1998, p. 366). Manufactures are investing increasingly in 'future gazing' to anticipate the products and services that will be relevant to people in the future. Indeed, some organisations employ futurologists specifically to track and advise on social and technological trends. Often future gazing involves the production of a product or service concept, which potential users can comment on. Focus groups are a favourite tool for gathering these comments. However, their suitability for this purpose remains questionable. According to Ireland and Johnson (1995):

> Most people cannot imagine themselves acting habitually in ways other than they currently do. In 1980, few executives could picture themselves at a keyboard any more than a fashionable girl of 1955 could imagine herself well dressed without bouffant petticoats...executives can tell you...about their office practices; what's wrong with airline seats; and why they buy certain brands of vodka and not others. They cannot tell with any accuracy how they will react *in the future* to a softly curved widget that electronically routes email, organizes meetings, and automatically drafts reports. (p. 58)

Norman (1998) goes on:

> Focus groups are fun. They provide a lot of information. Any interaction with potential customers is always informative. But focus groups can be very misleading. They tend to reveal what is relevant at the moment, not about what might happen in the future. Users have great trouble imagining how they may use new products, and when it comes to entirely new product categories – forget it. (p. 192)

This chapter details a scenario-based focus group procedure that addresses these problems.

9.2 PROCEDURAL APPROACH

The Oxford English Dictionary defines a scenario as a postulated sequence of future events. Gould *et al.*'s (1987 in Preece, *et al.*, 1994) development of the Olympic Message System (OMS) provides an early example of how scenarios can be used formally for discussing products. The development of the OMS proceeded by preparing printed scenarios of the user interface, which were used as a touchstone for discussing the systems acceptability, both amongst the design team members and with potential users (Olympic athletes, their families and friends). Figure 9.1 shows what such a scenario might look like for a more familiar system, an ATM (Automatic Telling Machine, or bank machine).

User:	- Inserts card
ATM:	Please enter PIN.
User:	- Enters 1 2 3 4
ATM:	(System validates PIN) Please select a service.
User:	- Selects 'Cash withdrawal'
ATM:	Would you like a receipt?
User:	- Selects 'No'
ATM:	Your money is being counted – Please remove your card.
User:	- Removes card
ATM:	Please collect your money.
User:	- Collects money

Figure 9.1 An ATM scenario

More recently Chin *et al.* (1997) used a scenario-based procedure to discuss the requirements associated with collaborative learning tools. Their procedure involves four steps:

Scenario generation. Ask potential users to develop a set of usage scenarios, each one exemplifying a distinctive and typical usage situation for the currently employed product. For example, Chin *et al.* collected a series of videotaped observations of particular classroom activities.

Claims analysis. Ask the potential users to identify features of these situations that they consider significant. Ask the participants to elaborate both what is good or desirable and what is bad or undesirable about the characteristics identified. Chin *et al.* extracted the characteristic 'large class size' to use the example above again. Some of the *pros* associated with large group size were: (i) 'may provide greater input and knowledge than smaller groups'; (ii) 'requires less grading than by the teacher if group is graded as a whole'. Some of the *cons* associated with large group size were: (i) 'not everyone in the group may be engaged'; (ii) 'demands greater accessibility to the equipment since the equipment must be shared among a larger number of students' (p. 166).

Feature envisionment. Ask the potential users to envision new features that would emphasise the pros and de-emphasise the cons.

Scenario envisionment. Ask the potential users to consider how the envisioned feature would modify the original scenarios and then repeat the claims analysis process.

This procedure identifies design requirements by envisioning new features. However, it may well have a role further downstream, evaluating new features associated with product and service concepts.

Norman (1998) argues that participants in focus groups have great trouble imagining how they may use new products because much of people's current behaviour is not conscious, 'we are primarily aware of our acts when they go wrong or when we have difficulty' (p. 192). The procedure of Chin *et al.* (1997) encourages participants to examine how they use existing products and services. It attempts to raise their awareness of their everyday acts by asking them to make these acts explicit. The requirement to produce usage scenarios forces the participants to think about what they do, when they do it and how.

Moreover, the procedure encourages participants to acknowledge that the products and services that support these acts are not ideal. One reason why people have difficulty anticipating the potential benefits of a new concept is that they cannot see any problems with existing products and services. Chin *et al.*'s procedure makes the impact of envisioned features definite by requiring participants to incorporate these features into their original usage scenarios and consider how they would be modified. Equally it would be possible to incorporate new features of product or service concepts into usage scenarios and request participants to evaluate their positive or negative impact.

Significantly, the procedure also encourages participants to examine their existing situations in a way that is informal and consistent with how they naturally discuss these things. For example, Lloyd (2000) notes the spontaneous and informal use of scenarios by engineers who would tell stories to each other to develop a common understanding of a particular situation. Later the situation could be referred to in terms of the 'Edwards fiasco' or the 'Lift edge problem'. These stories would 'encapsulate personal experience and contribute to a larger narrative at either team, project, departmental, or organisational level' (Lloyd, 2000, p. 368).

Such story references litter conversations within design teams. Indeed, people without any design experience also seem to use scenarios informally when discussing systems. For example, Evans (1998) recounts the tale of the fake ATM (Automatic Teller Machine, or bank machine) in reviewing the considerable urban mythology associated with cash machines. The story goes that a gang of thieves sites a dummy ATM in a shopping centre. As people unsuccessfully attempt to withdraw money the fake ATM records their personal identity numbers and card details. The gang then uses this information to access their accounts. In this scenario, the validity of the claim (i.e. is this technically possible) may be less important than the opportunity to discuss the security risks posed by ATMs.

In a similar vein, Payne (1991) discusses the spontaneous generation of scenarios by participants asked to explain how ATMs work. One interesting observation from Payne's work is that:

> Subjects freely mix explanation types using, perhaps, design-rationale in one breath and direct experience in the next, they juxtapose pre-existing conceptions with novel constructions, and they summon special-purpose analogies to deal with small components of the total systems. (p. 19)

In other words, scenario generation can be seen as part of people's broader ability to present explanatory models through the telling of stories that mix together many different forms of description. Scenarios thus provide a naturalistic means by which participants can examine how they use existing products and services.

The following case study illustrates a scenario-based focus group that follows Chin *et al.*'s (1997) procedure. However, rather than asking participants to envision features for a new product this study asks them to evaluate the features of a product concept.

9.3 CASE STUDY

A number of organisations are currently engaged in the creation of electronic wallets, which typically combine the traditional wallet with computer and, in some instances, communication technology. Their rationale is that the wallet is inherently associated with personal finance and will therefore provide a natural platform for electronic commerce and banking (Cooper *et al.*, 1999). For example, AT&T's electronic wallet, the Mondex wallet, is intended to replace the traditional wallet within a smart card system. Smart cards are similar to conventional credit cards but contain an embedded computer chip that can store monetary value and information. AT&T's wallet allows users to perform online transactions and view the balance and transaction details that are stored on their smart cards. The objective of the following case study was to evaluate a particular electronic wallet prototype.

9.3.1 Participants

The participants were 24 adults aged between 21-36. They were selected on the basis of having experience with a variety of personal technologies (e.g. mobile phone, personal digital assistant) and not being antagonistic towards banking systems. These constraints were placed on selection in order to facilitate discussion. Four groups, two all male, two all female, each consisting of six participants were conducted. Single sex groups were used to prevent 'male experts' potentially dominating mixed sex discussions.

9.3.2 Procedure

The participants were asked to develop scenarios exemplifying how they currently use their wallets and banking systems such as ATMs and Internet banking products. These products were chosen as they were regarded as being most closely associated with the prototype. They were given pens, pencils and paper and were instructed to illustrate their usage scenarios. The participants were asked to identify features of these situations that they considered significant. They were then asked to elaborate both what is good or desirable (the pros) and what is bad or undesirable (the cons) about these features as a group. Finally, they were shown an electronic wallet prototype and instructed to evaluate how the features of the new technology would emphasise the pros or de-emphasise the cons identified earlier.

9.3.3 Results and Discussion

Two of the features identified by the participants are reported below.

Storage

Many of the participants (75 per cent, 18 of the 24) identified storage as an important feature of wallet use. For example, that 'you can store most any small thing in a wallet' was regarded as a pro. However, that 'you can store only a limited amount of small things in a wallet/purse' was regarded as a con. The perceived inability of the prototype electronic wallet to store objects other than a few banknotes was anticipated to be a real problem. Whilst the participants most closely associated the wallet with the action of storing financial objects, such as cash and bank cards, they also revealed in their scenarios that they used it to stash a multitude of other objects. In fact the participants, prompted by the storage scenario in Figure 9.2, discussed hoarding rather unusual items including plasters, textile samples, and teeth! Furthermore, several of the participants went on to say that they only really carried a wallet to store sentimental things, such as pictures of their children. Indeed as the discussion moved away from talking about wallets as financial artefacts it became obvious that they were as much an emotive artefact as anything else.

Scenario 1:

I was at a party and bumped into a friend. He gave me a newspaper cutting, which he thought I might be interested in. I was a bit drunk so I stuck it in my purse till I got home.

Figure 9.2 Step 1: scenario generation

Access

The prototype faired better with regard to access. The majority of the participants (83 per cent, 20 of the 24) identified accessibility as an important feature of banking systems. Although the only constraint placed on the participants was to consider a 'banking system', all of the participants produced ATM usage scenarios. For example, that 'you can use an ATM at anytime' to recover information was regarded extremely positively. However, that 'you have to know where an ATM is located (sometimes in an unfamiliar place)' was regarded very negatively. As previously, the participants used these and other

ATM pros and cons to evaluate the features of the concept electronic wallet prototype that they were presented with. Clearly the portability of the electronic wallet maintains people's ability to access financial information anytime. Moreover, the electronic wallet would allow them in the future to access financial information 'anyplace'. Furthermore, the portability of the concept electronic wallet meant that they did not have to search for the system beyond their bag or pocket.

9.4 CONCLUSIONS

There is nothing new in people using narrative to guide and inform others. Barthes (1977) argues that people in every age, in every place and in every society have used stories to share their experiences. Fiske (1987) further emphasises the significance of narrative by comparing it to language as a 'fundamental sense-making mechanism'. It is hardly surprising then that people are currently using narrative to guide and inform others about the design of future products and services.

The narratives or *scenarios* in the case study performed a number of roles to this end. Firstly, they forced the participants to consider their existing behaviour:

> The scenarios enable the users to immediately connect the scenarios to their own image of their work. Users develop a greater sense of ownership of the analysis and design because the activities are grounded in authentic activities occurring in the user's world.
>
> (Chin *et al.*, 1997, p. 166)

This provided a familiar context, which the participants owned, and by which the participants were better able to decide on the utility of the prototype. Furthermore, Chin *et al.*'s procedure itself is consistent with how people make decisions. They make the argument that their procedure is just an incrementally more systematic variant of the decision process that people engage in day-to-day:

> When people shop, they perform claims analysis as they assess the features of the products: how will this automobile do in the ice scenario, are the tires wide enough for the beach? Claims analysis merely asks that the reasons why a specific feature is good or bad be explicitly elaborated.
>
> (Chin *et al.*, 1997, p. 166)

Indeed, the participants in the case study performed the claims analysis procedure with ease. Nevertheless, the whole procedure proved quite time-consuming. This was one of the few problems experienced with the methodology.

Secondly, the scenarios also restricted the scope of the envisionment task. This possibly enhanced the participants' ability to evaluate the prototype. The idea that constraints can enhance this activity at first seems paradoxical particularly given the widely held belief that such activities require as few restraints as possible. Macaulay (1996), for example, argues that the purpose of a focus group is to allow a group of users to talk in a free-form discussion with each other. However, concentrating only on a small subset of problematic situations may in fact enhance participants' ability to envision future situations. Indeed, Perkins (1981) indicates that a person is better able to notice or recognise the unexpected by thinking within a constraining framework. Furthermore, Finke (1992) showed that in a design task the relative number of creative products envisioned increased significantly as the task became more highly constrained.

Finally, and perhaps most importantly, the scenarios provided a communicative tool by which people could discuss the prototype. Potential users are becoming increasingly involved in design. Participatory design, for example, advocates a heavy user involvement in system design, in which they actively engage in designing the systems that they will eventually use (Macaulay, 1996). Scenarios seem to allow people from different backgrounds to discuss systems by providing a common and perhaps natural sense-making mechanism. Indeed, observations of the formal and informal use of scenarios in discussing product development have prompted some to conclude that design cannot proceed without some notion of narrative (Carroll, 2000).

Significantly, the scenarios allowed the participants to communicate as a group. Lunt and Livingstone (1996) note that such public discussion is particularly important in the production and reproduction of meanings in everyday life. They go on to argue that focus groups could:

> ...be understood not by analogy to the survey, as a convenient aggregate of individual opinion, but as a simulation of these routine but relatively inaccessible communicative contexts that can help us discover the processes by which meaning is socially constructed through everyday talk.
>
> (Lunt and Livingstone, 1996, p. 85)

In their opinion the former view of focus groups predominates contemporary practice. As such, scenarios may be key in establishing the focus group as a theoretically significant approach rather than as a convenient (or contaminated) source of singular belief.

In this chapter, focus groups have been used to generate design concepts. Rather than allowing free-wheeling discussion, we have taken care to exercise some constraints on participants; both in terms of the topics under discussion (as is often advised in the running of focus groups) and in the mode of responding. The use of pencil and paper to allow people to clarify, sketch and note their ideas, in addition to purely verbal discussion has proved a useful and, for most participants, enjoyable extension of the focus group methodology. Conducting discussions around scenarios was seen as a further enhancement of the method and provided the researchers with many useful insights into the everyday activity of the participants. Indeed, subsequent field observations failed to generate data that went beyond the scenarios generated in the focus groups.

Consequently, it is proposed that 'focused groups', in which the discussion of technology use is concentrated on the development and sharing of scenarios, and in which participants are provided with means of expressing their ideas over and above merely verbally, can offer many benefits to designers:

- scenarios can be participatively designed, reflecting both the experiences and aspirations of participants and also providing stimulation for subsequent concept evaluation;
- sketching scenarios provides an opportunity for people to express ideas and feelings that they might not, at first, verbalise;
- restricting the scope of discussion can often liberate creativity and lead to novel and exciting proposals.

9.5 REFERENCES

Barthes, R., 1977, *Image-music-text*, (London: Fontana).

Carroll, J. M., 2000, *Making use: Scenario-based design of human-computer interaction*, (Cambridge, MA: MIT Press).

Chin, G., Rosson, M. and Carroll, J., 1997, Participatory Analysis: Shared Development of Requirements from Scenarios. In *CHI 1997*, (New York, NY: ACM Press).
Cooper, L., Johnson, G. and Baber, C., 1999, A run on Sterling – personal finance on the move. In *Third International Symposium on Wearable Computers*, (Los Alamitos, CA: IEEE Computer Society).
Evans, H., 1998, ATMs. www.crd.rca.ac.uk/~helen/report/index.htm (accessed 20 March 2000).
Fiske, J., 1987, *Television culture*, (London: Routledge).
Finke, R., 1992, *Creative cognition: Theory, research and applications*, (Cambridge, MA: MIT Press).
Ireland, C. and Johnson, B., 1995, Exploring the future in the present. *Design Management Journal*, **6**, pp. 57-64.
Lloyd, P., 2000, Storytelling and the development of discourse in the engineering design process. *Design Studies*, **21**, pp. 357-373.
Lunt, P. and Livingstone, S., 1996, Rethinking the focus group in media and communications research. *Journal of Communication*, **46**, pp. 79-98.
Macaulay, L., 1996, *Requirements Engineering*, (London: Springer).
McDermott, C., 1998, *20th Century Design*, (London: Carlton Books).
Norman, D., 1998, *The invisible computer*, (Cambridge, MA: MIT Press).
Payne, S.J., 1991, A descriptive study of mental models. *Behaviour and Information Technology,* **10**, pp. 3-22.
Perkins, D., 1981, *The mind's best work*, (Cambridge, MA: Harvard University Press).
Preece, J., Rogers, Y., Sharp, H., Benyon, D., Holland, S. and Carey, T., 1994, *Human-Computer Interaction*, (Harrow: Addison-Wesley).

CHAPTER 10

Harnessing People's Creativity: Ideation and Expression through Visual Communication

Elizabeth B.-N. Sanders and Colin T. William

10.1 INTRODUCTION

The people who buy and use the products we create are not typically invited to play in the fuzzy front-end of the development process because it is commonly believed that they are not creative. But participation early in the front-end is needed to drive truly human-centred product development.

We have learned how to harness the creativity of potential end-users very early in the development process. The research methods in this chapter explore not only what people say and do, but also what people make. The *say* methods are rooted in verbal communication and are used in situations such as traditional focus groups. Methods such as applied ethnography focus on what people *do* through direct or indirect observation. The *make* methods are unique in their ability to elicit creative expression from everyday people.

This chapter describes and shows examples of 'make' toolkits we use to elicit creative thinking from people who will buy and use future products and services. It also outlines how all the methods (say and do and make) are used together in order to fully understand people's past, present as well as anticipated experiences with products and services.

10.2 THE CURRENT SITUATION

There is a widely held belief shared today by professionals involved in new product and service development that consumers are not creative. This belief has many variations including:

> 'Consumers do not know and cannot express their needs and/or dreams.'
>
> 'Consumers cannot imagine or envision how their future could be different from the present.'
>
> 'Consumers cannot come up with ideas for new products or services to improve their lives.'
>
> 'Consumers cannot even recognise good ideas that are put in front of them in the form of concepts or prototypes.'

This belief has come from many years of market research dominance in the product development arena. Traditional forms of market research rely on verbal communication between people, i.e. on what people *say*. The 'say' methods are useful for getting an idea of what consumers can tell you and want to tell you in words, but they are very limited in the context of all the other methods that can be used to understand people's unmet needs and dreams.

Even the language that market researchers use to describe the people we design for, such as *consumers, users* and *customers*, puts them into roles with limited and very clearly defined boundaries. 'Consumers are people who shop.' 'Users are people who use products and services.' 'Customers are purchasers.' By putting people in narrow categories, we limit their ability to contribute creatively.

As a result, the people who will be buying and using the products and services we create have not been invited to play in the front-end of the development process. At best, they are invited to participate at the concept evaluation phase or in usability testing. Participation in the middle or end of the design development process is not enough to drive truly human-centred product and service development. To do so requires that we find ways to harness the creativity of ordinary people early in the design development process.

10.3 THE EMERGING POINT OF VIEW

We have found that there are research methods and tools that everyday people can use to contribute in the front-end of the process. We have seen, contrary to the prevailing point of view, that everyday people are creative when given appropriate 'tools'. We can harness their creativity early in the development process.

Examples of the innate creativity of everyday people are becoming more evident daily. Desktop publishing by people not trained in design was the first sign. Now personal websites are proliferating, as are websites where people can, for example, design their own shoes or decorate their own t-shirts. On the non-electronic side, we see the rapid growth of activities such as scrapbooking and a renewed interest in folk art.

With the advent of new forms of electronic communication, people today have become more demanding 'consumers'. They are in the position now to be participants, even innovators, in the new product and service design and development process.

10.4 THE OPPORTUNITY: HARNESSING PEOPLE'S CREATIVITY

Before talking about how we have learned to harness the creativity of everyday people, we need to define what we mean by creativity. We take a transdisciplinary view:

> Creative thinking in all fields occurs preverbally, before logic or linguistics comes into play, manifesting itself through emotions, intuitions, images and bodily feelings. The resulting ideas can be translated into one or more formal systems of communication such as words, equations, pictures or music or dance only after they are sufficiently developed in their prelogical forms.
>
> (Root-Bernstein and Root-Bernstein, 1999)

We will use *ideation* to refer to the preverbal idea stage and *expression* to refer to the translation of those ideas into formal systems of communication.

According to Koestler (1964), every creative act involves *bisociation*, a process in which previously unrelated ideas are brought together and combined. He contrasts bisociation with association. Association refers to previously established connections amongst ideas. Bisociation involves making entirely new connections. According to

Koestler, bisociation only occurs when the person has been thoroughly involved in the problem or situation for a long time. Koestler also emphasises the importance of dreams, in that dreaming involves bisociation at an unconscious level.

To harness people's creativity we need to support both ideation and expression. Here is the four-step framework we use:

1. Immersion.
2. Activation of feelings and memories.
3. Dreaming.
4. Bisociation and expression.

The *immersion* step lasts from one to several weeks and takes place in the natural context of the experience being studied, usually the participant's home or place of work. We guide the participants in daily self-documentation of their thoughts, feelings and ideas about the experience being investigated. We may also ask them to observe and document their current behaviour as well.

The next three steps take place in face-to-face meetings, either individually or in groups. First we engage the participants in one or more exercises with 'toolkits' that have been designed to evoke and *activate related memories and feelings*. The research tools we use for the activation exercises are predominantly visual, and are discussed in greater depth later in this chapter.

Then we invite them to participate in an exercise with a toolkit designed to encourage their *dreaming* about their future or an ideal experience. Again, the research tools are predominantly visual.

In the final *bisociation and expression* step we invite them to bring expression to the ideas they are having. This is the most exciting part – we have seen people express completely new ideas for products or services within a few minutes. It is as though they are ready to explode with ideas. The exercises/toolkits that we use for this step are deliberately abstract and ambiguous.

10.5 DIFFERENT APPROACHES TO LEARNING: WHAT PEOPLE SAY, DO AND MAKE

The research methods in this chapter explore not only what people say, but also what people do and make. Typical approaches to learning about people focus on what they say and do. As mentioned above, the 'say' methods are rooted in verbal communication, such as through traditional focus groups. This can generate a lot of information, but it is limited to what people are able to put into words. Methods such as ethnography focus on what people *do*, through direct observation or through logs and diaries. Such methods also provide much information about behaviours, but they often lack in accessing people's underlying motivations or emotions.

The 'make' methods described in the next section look at what people *make* with the creativity-based research tools you give them. 'Make' methods enable creative expression by giving people ambiguous visual stimuli to work with. Being ambiguous, these stimuli can be interpreted in different ways, and can activate different memories and feelings in different people. The visual nature liberates people's creativity from the boundaries of what they can state in words. Together, the ambiguity and the visual nature of these tools allow people much room for creativity, both in expressing their current experiences and ideas and in generating new ideas.

10.6 EXAMPLES OF CREATIVITY-BASED RESEARCH TOOLS

This section provides examples of research tools that allow participants to use their natural creativity to express their ideas. These include pre-meeting tools for use before a group session to facilitate immersion, tools for activating memories and expressing emotions, tools for mapping processes and routines, and tools for bisociation and expression of ideas and ideals. The array of possible tools is limitless, and you should not constrain yourself to those outlined in this chapter. Experiential research tools are not 'one size fits all', and you should develop tools for each project with that in mind.

10.6.1 Pre-Meeting Immersion Tools

Mailing participants some tools in advance of a group meeting can enhance the quality of the meeting that takes place. Participants can think about issues to be discussed and about their own experiences. Below we highlight three pre-meeting tools, which can be used individually or together in one booklet.

Workbooks

A workbook can contain different types of questions, such as demographic information, opinions and information about things people own or use. This lets you learn details about the people without using up meeting time, and thus is suited to less discussion-based issues.

Hands-on exercises also can be included; these can be completed and brought to a meeting, or mailed back to you in advance (the latter allows you to develop more relevant follow-up questions). Assigning exercises ahead of time in workbooks can save time at the meeting, but it is important that you give enough workspace and clear instructions. Also, make exercises as interesting as possible to increase the chance of people completing them.

Diaries/day-in-the-life exercises

These methods give people space in a workbook to record and become immersed in experiences. This can attune them to details about activities they might normally take for granted. For instance, a diary of commuting experiences might have people record the amount of time spent in the car each day, amount of time spent stopped in traffic, and how they felt and occupied themselves during commuting downtime.

In a day-in-the-life exercise, people outline their typical day, either specific aspects (e.g. use of computers for work and leisure, workspaces etc.) or across the board. This exercise can be useful in identifying opportunities for new ideas, as people can explain their experiences and the emotions attached to them.

Send a camera home

Giving people a camera to document their experiences can enhance both workbooks and diaries. They take pictures to document specific aspects of their lives, and mount them on workbook/diary pages to show what they do. This adds a level of richness to text-based methods – as the saying goes, a picture really can be worth a thousand words. So, for instance, a picture of a workspace can give you deeper insight into the ways people organise their lives and their stuff.

Two options for this are the disposable camera and the Polaroid camera. The former is cheaper, with minimal loss if a participant is dropped from the study. One downside is that participants must leave time to get film developed. Polaroid cameras and film are much more expensive, but they allow people to take and evaluate pictures on the spot. Further, people can immediately annotate the picture.

10.6.2 Tools for the Group Meeting

Below are just a few examples of tools that can be used within group meetings. These tools can also be used individually. They are designed to allow participants to use their previous immersion to stimulate creative exploration. Again, these are only examples to which you should not feel constrained.

Collages

Collaging allows people to articulate experiences through pictures and words. Collaging involves giving people a set of picture and word stickers and a space on which to arrange them according to your instructions (see Figure 10.1).

Figure 10.1 A sample collage set, and people creating collages in a group

Collaging is ideal for activating feelings and memories, as well as for dreaming. The stimulus set should not be too precisely defined – some stimuli should be ambiguous enough in meaning to allow participants to impose their own interpretation to facilitate creative expression. A collage exercise for our commuting project might give people a workspace divided into two halves, with one for expression of emotions surrounding current commuting experiences, and the other for dreaming about the emotions of commuting in the future.

There are four steps to developing an effective collage toolkit: brainstorming, pilot testing, refinement and production.

Brainstorming: In brainstorming ideas for a collage toolkit, the goal is to provide the participants with the means to communicate across a wide array of experiences. The challenge, then, is in selecting stimuli – words and pictures – which will not only cover experiences you might anticipate, but also others that you will not anticipate. For this reason, the stimuli must be intentionally ambiguous, so as to allow the participants to impose their own meanings. Moreover, you should involve as many people as you can in your brainstorming sessions, to expand the horizons beyond your own experiences.

The brainstorming session will focus on selecting words and pictures for use in the toolkit. Words are the easier part – anyone can brainstorm words that reflect their experiences. Some sample words for our commuting project might include 'traffic', 'thirsty', 'waiting', 'music' or 'boring'. In fact, you should ask at least three people to brainstorm words in order to get a good variety.

Words by themselves, however, are fairly narrow in scope, and that is what leads to the importance of pictures. Brainstorming an image set will be crucial in enticing your participants to communicate. Images can come from anywhere, e.g. magazines or personal photographs. What is important is creating a set that will be evocative. A picture of a person in a car is relevant to our project, but it will not necessarily evoke discussion or emotions. On the other hand, images such as a picture of a traffic light, a cup of coffee, or a jagged speech bubble filled with '@#$%&*!', will be more potent, more real. But at the same time, you should include images you may not imagine to be part of the experience. Including a picture of a serene scene might evoke someone to talk about their ideal, stress-free commute. So the idea is to cover as broad a range of emotions and experiences as one can, even if they may not fit the expected model of the experience.

On the practical side, you should already be anticipating how you will present these images and words to people. If you plan on giving people sticker sheets of pictures and words, then plan out what size stickers you will use, so that you can anticipate how large or small your images need to be. One image may seem perfect but when reduced down to sticker size, it may not be intelligible. On the other hand, if you choose to use paper sheets of images, you can use larger or nonstandard shapes of images (see the section below on 'Production' for more on this). Here are a few rules of thumb for what to include in a collage set:

- a balance of positive and negative images and words;
- a balance of abstract and concrete images and words;
- both natural as well as man-made things;
- people of all types – male and female, young and old, racial balance etc.

Once a large set of ideas has been brainstormed, the set will inevitably need to be narrowed down. Overwhelming participants with an immense toolkit is not recommended. Reduce the set to a manageable level, certainly no more than 100 words and 100 pictures, and preferably 150 of the two combined. Distilling your set to manageable proportions will be a challenge, but it will be helped if you can spot redundancies. In our example, a picture of a clogged highway and the words 'traffic jam', both address the same thing, so it may be better to discard one or the other. At this stage you may not need to narrow down your set all the way. If you get rid of the most obvious redundancies, then you can solicit help in the next stage to narrow the set further.

Pilot Testing: It is imperative to try out your instructions and picture/word set on a few pilot subjects to get their feedback. Whilst your instructions may seem obvious to you on paper, they may seem vague or unclear when spoken aloud to a person unfamiliar with the task. Remember, people are not asked to do this kind of task everyday, so it is important to make sure in advance that they understand it.

During a pilot test you will run a few people through your task, preferably people who do not know about your project. Within your pilot test, you should try to approximate your test conditions as much as possible. So if your target is to complete a task in 20 minutes, then you should try to do so in your pilot test also.

Where a pilot test may deviate from your plan is in soliciting feedback. When the pilot test people are trying to build their collage, you can ask them about what they are thinking – are there any concepts they would like to express, but for which they cannot find an image or word? Any images or words they think are redundant? This can be done as they look, when their thoughts are fresh, or after they are finished. The latter will allow you to deviate less from your test plan.

Pilot testing can be invaluable in finishing out your toolkit. Almost always it will lead you to make subtle changes to your instructions. Also, you will get a clearer idea of the completeness of your stimulus set. You should be prepared to do a few pilot tests, and to change instructions and toolkits on the fly as you conduct them, especially if you are just beginning to learn these new tools and methods.

Refinement: Once you have run your pilot tests, you should make refinements to your toolkit. Narrow down the image/word set again, if it is not already at a manageable level. If you make major changes, do not be afraid to run another pilot session. They seldom take much time, and the enhancements to your toolkit can be major.

Production: Once your toolkit is finalised, you can proceed with production. You should have already given consideration to how you want to present your images to people, and there are costs and benefits either way.

Sticker sheets allow participants to work more quickly, and are less cumbersome; however, sticker sheets can be expensive, and they place strict limits on image sizes. Moreover, if you choose small stickers, people observing the groups through a one-way mirror may be unable to see the images as participants talk about what they have created.

The other option is to print your images on sheets of paper, and have participants use scissors to cut them out and glue sticks to paste them down. This method is less expensive, and it allows more freedom with image sizes; however, it also requires more time and effort for participants to create a collage this way. In producing your collage toolkit, you should make several extra copies to give to those observing the session. Be sure to have enough supplies (e.g. scissors, glue, tape and coloured markers) for everyone.

Figure 10.2 A sample cognitive mapping kit and a completed cognitive map. A complex structure like the one on the right would be difficult to express in words

Cognitive mapping

Cognitive mapping allows people to map out processes and events, or their understanding of categories or systems. Cognitive mapping toolkits usually contain a poster board and symbolic shapes, which can be used to map out connections, clusters or hierarchies of concepts (see Figure 10.2). Words may also be included. These tasks let people express and develop bisociations in thought processes and structures. The toolkit provides people a medium that channels and simplifies their ability to express complex concepts.

The general process of developing a cognitive mapping toolkit is much the same as that for developing a collage toolkit. That is, you follow the same brainstorming > pilot testing > refinement > production procedure. The major differences lie in anticipating the workspace people will need and the kind of shapes that will help them express their ideas. The shapes in Figure 10.2 include a house, a cloud, sun shapes, circles and hands. As mentioned above, the bright orange and yellow shapes might be used to reflect positive emotions, whereas a grey cloud may be very negative. The shapes may be used literally or they may be used metaphorically. It is advisable to include very simple shapes such that many different interpretations are possible for each shape.

For the development of your mapping toolkit, it can be useful to go to an art supply or a hobby/craft store and buy a broad sample of the coloured shapes available. You will probably also need to produce some shapes of our own. Your brainstorming with colleagues will help to put together the appropriate selection. You can use your pilot sessions to get feedback, and to see which shapes people use to express themselves.

In selecting word stickers, the procedure will be essentially the same as for developing a collage toolkit; however, since cognitive mapping is better used for addressing experiences at a more specific level, you will want to make your selections more concrete (i.e. less ambiguous) than for an emotion-centred collage. This will allow you to start pulling out specific areas of opportunity in looking at what people are doing, thinking and feeling today.

As with the collage toolkit, be prepared with extra copies of your toolkit for your audience.

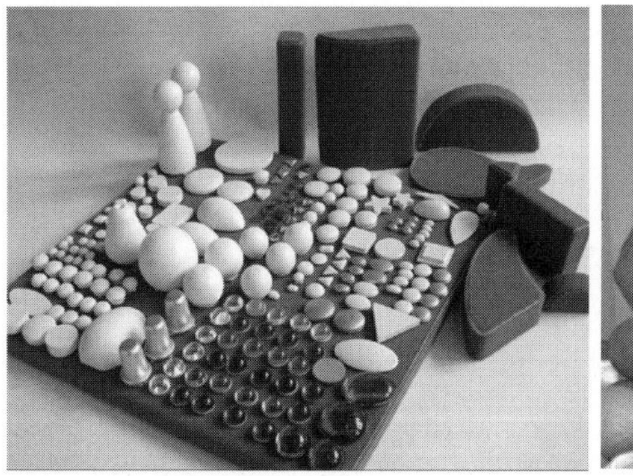

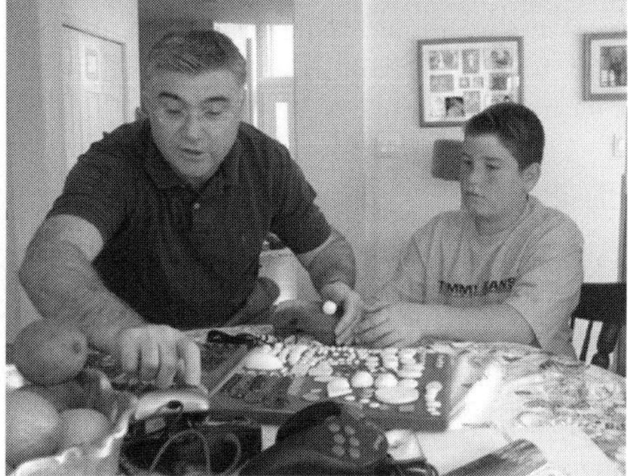

Figure 10.3 A sample Velcro-modelling kit, and a father and son collaboratively creating an ideal model using this kit

Velcro-modelling

Velcro-modelling allows people to embody and express their ideas in low-fidelity, three-dimensional models. Ideal for use after other tasks have been used for immersion, activation and dreaming, Velcro-modelling allows people to actively embody their ideas in a hands-on manner. A Velcro-modelling kit will consist of shapes, buttons and other items (see Figure 10.3) that are ambiguous in purpose, maximising the opportunity for people to imagine and impose their own thoughts in the expression of their ideas. In this case, however, the basic 'building blocks' (i.e. the larger dark grey components in the upper right corner of the kit photograph) are three-dimensional forms covered with material to which Velcro will stick. The smaller control elements (i.e. the white and the coloured button-like shapes in the same photo) are three-dimensional as well. Some are purchased and others are custom made. All the smaller control elements have Velcro on the back so they can stick to the grey forms. We also supply extra Velcro dots so that people can stick the grey forms to one another.

For our commuting example, we might create a model cockpit for a car with surfaces coated in Velcro. Participants could arrange specific controls or features according to their needs and to the needs of their passengers, to create and describe the ideal cockpit for the everyday commuter.

It is not possible to cover the creation of a Velcro-modelling toolkit in this short chapter. Most pieces are custom-designed and the collection of a full set may take years. We started with a small set and still find the need to add on to it every time it is used. Contrary to the toolkit sizes prescribed for image collaging and cognitive mapping, it is best to have very large Velcro-modelling toolkits. For example, for a group of four or five people, we might use hundreds of grey forms of many shapes and sizes together with hundreds of smaller components. At the point at which Velcro-modelling is used, people usually have ideas in mind and they quickly scan through hundreds of components to find those that best fit their ideas. A Velcro-modelling toolkit for an individual would be smaller, but still offer plenty of choice in components so that the person could quickly give form to any ideas that come to mind.

Velcro-modelling is a very powerful tool for allowing ordinary people to express their dreams. It is often the case that people start grabbing pieces and using them intuitively before we can even finish giving the instructions.

10.7 GUIDELINES FOR USING TOOLS

10.7.1 Selecting Tools

With so many tools available, it is important to develop a plan for using them. Getting the most out of tools requires having defined goals about what you want to learn, and a procedure that builds towards those goals. In developing your methodology, you should think about your goal, then pick tools that will facilitate people's thinking and talking about their experiences. In doing this, consider the following issues.

Which research tool or tools will access what you want to learn? Consider the aspects of experience you wish to access. As discussed in the preceding section, different tools are better for accessing cognitive versus emotional aspects of experience. Certain tools can access people's ideal experiences and ways for them to realise those ideals. Your goals will help dictate the types of tools you will use.

What do you want out of the tools? Different tools yield different results. Some are more useful for the discussion they generate. For instance, collages enable storytelling, as participants talk about their collages. On the other hand, with Velcro-models the artefacts have greater importance. These artefacts may be a primary goal, for example, to serve as inspiration for designers. But even then, the discussion of the models can be extensive and can serve as an important source of inspiration.

In what order should the tools be presented? This issue is very important to the results you will get. Often the results of a tool will be incomplete unless participants have thought about the experience situation being investigated in advance. You can handle this by sending a workbook and camera to each participant before the meeting. This allows them to pay more attention to everyday life experiences, priming them for participation in the group discussion. Similarly, having participants create a past-present-future collage will prime them for thinking about specific ideal experiences or products.

A good approach to deciding on the order is to follow the four-step framework described previously. Work from the general to the specific – start with current experience, explore it in greater depth, then use this as a baseline for evaluating and/or creating new ideas. Below is an example of tool ordering for our hypothetical study of commuting experiences for exploring automobile cockpit design.

1. Immerse people in their everyday experiences of commuting via a workbook and camera.
2. Open the group meeting with discussion of things they noticed whilst filling out the workbook.
3. Have them create collages about the emotional aspects of commuting today versus in the future in order to activate their feelings and memories.
4. Have them create a cognitive map of the ideal experience of commuting to encourage their dreaming.
5. Have them work with a Velcro-model car cockpit to support bisociation and expression. For example, they could remove elements that inhibit their ideal experience and add elements that would enhance the experience.

10.7.2 The Discussion Guide

Having created a general order for using tools, you can now draft a discussion guide, the formal instructions for running the meeting. This should cover the order of events, specific instructions for each part of the meeting, amount of time for each exercise, goals for each tool and follow-up questions. This guide is useful for standardisation, both from session to session and across different people who might be running sessions.

In writing your discussion guide, consider your goals and the time available. It is no small challenge to learn about people's experiences in two hours whilst giving all participants the opportunity to speak. Outline specific instructions for your tools and estimate the time needed for each. Be generous and realistic about this: estimate time for participants to introduce themselves, to discuss any homework you gave in advance, to carry out each exercise, and time for each person to talk about what they create in each exercise.

Next you should run a pilot session using the discussion guide and prototype tools. Recruit volunteers and run them through. Keep track of the time for each exercise and discussion, and get feedback from your volunteers on timing issues (too much time or too little for each exercise?) and instructions (clear, unclear?).

Based on these sessions, make changes to the guide. When estimating time, remember that in a group you will have more people than in your pilot session. If you run too long, consider putting certain exercises into the workbook for participants to do in advance to bring with them to the meeting. If you still run too long, explore ways to meet your goals using fewer tools.

You also can change instructions that were unclear or did not meet your goals. Good feedback from pilot subjects can lead to changes to both the instructions and the tools themselves, such as suggestions on stimuli that should be added to the exercises. Seeing the tools in action will provide you with insight for refining them for better discussion and results.

Having modified your procedure, test the full discussion guide again for time, and test any modifications to instructions that you make. After a few rounds of testing and iteration you will be ready to go.

10.8 RESULTS OF CREATIVITY-BASED RESEARCH

Just as the toolkits support the creativity of everyday people, the artefacts people make from the toolkits support the creativity of designers. We often find that the design team members keep the artefacts such as collages, cognitive maps and Velcro-models to surround themselves with during their own ideation and expression activities.

There are methods for both qualitative and quantitative analysis of the data from the say, do and make approaches. The most basic methods will centre on identifying the items that occur most frequently on collages or in transcripts. Such items can serve as a starting point from which to go back to transcripts (i.e. the written protocol of what people said in the group meetings) to look for recurring themes.

More powerful statistical methods include cluster analysis and the use of Pathfinder Associative Networks (Schvaneveldt, 1990); these can be useful for identifying patterns in data. A discussion of the analysis of data could fill a chapter or even a textbook to itself, however, so it is beyond the scope of this chapter.

10.9 CONCLUSIONS

To drive truly human-centred product and service development requires that we harness the creativity of ordinary people early in the development process. Although this chapter has focused on the role of the 'make' tools, it should be clear from the example that a carefully planned execution of all three categories of methods (say and do and make) is needed to develop a successful understanding of people's unmet needs and dreams. It is in the convergent territory that lies at the intersection of what people say and do and make that the useful insights can be found.

We have focused on the use of the make tools for a number of reasons. First, because they are new and because we have found that these tools are the most effective way we know to elicit the creativity of everyday people. Use of the make tools has the added benefit of resulting in artefacts such as collages, mappings and Velcro-models. Designers and other members of the development team can quickly become immersed in the minds and hearts of their potential end-users when they hear people talk about the artefacts they have created with the make tools.

The methods and tools described in this chapter have been used to drive and inspire the design development of products and services in many different industries: consumer, computer, medical and industrial. The Zip Drive by Iomega that was introduced in the mid 1990s is one of the better recognised results of this approach.

10.10 REFERENCES AND RELATED READINGS

Buzan, T. and Buzan, B., 2000, *The Mind Map Book* (Millennium Edition), (London: BBC Worldwide Ltd).

Capacchione, L., 2000, *Visioning: Ten Steps to Designing the Life of Your Dreams*, (New York: Penguin Putman).

Gardner, H., 1983, *Frames of Mind: The Theory of Multiple Intelligences*, (New York: Basic Books,).

Horn R.E., 1998, *Visual Language: Global Communication for the 21st Century*, (Bainbridge Island, Washington: MacroVU Incorporated).

Koestler, A., 1964, *The Act of Creation*, (London: Hutchinson & Co.).

McCloud, S., 1994, *Understanding Comics: The Invisible Art*, (New York: Harper Collins).

McKim, R.H., 1980, *Experiences in Visual Thinking*, (Wadsworth, California: Brooks/Cole).

Root-Bernstein, R. and Root-Bernstein, M., 1999, *Sparks of Genius: The 13 Thinking Tools of the World's Most Creative People*, (New York: Houghton-Mifflin).

Sanders, E. B.-N., 2000, Generative Tools for Co-designing. In *Collaborative Design. Proceedings of CoDesigning 2000*, Coventry, UK, 11-13 September. Scrivener, S., Ball, L.J. and Woodcock, A. (eds), (London: Springer-Verlag), pp. 3-7 (Keynote).

Schvaneveldt, R.W., 1990, *Pathfinder Associative Networks: Studies in Knowledge Organization*, (Norwood, New Jersey: Ablex Publishing).

CHAPTER 11

Beyond Focus Groups? The Use of Group Discussion Methods in Participatory Design

Joe Langford, John R. Wilson and Helen Haines

11.1 INTRODUCTION

By their very nature, focus groups are participatory processes, engaging people in discussion-based activities for a wide variety of purposes. Focus groups help researchers to understand users. They can lead to a more in-depth and empathic understanding of how and why people do things. They can be used to gain their views on the things they use, including what works well for them – and what does not.

In their simplest form, focus groups can be seen mainly as a way of gathering knowledge, which can then be used by experts such as designers and engineers. The experts, in effect, using the knowledge on behalf of the users. But is this using the full potential of the method? In participatory design terms, using information on behalf of the users is a relatively limited form of participation. It represents just one end of the spectrum of ways by which users and other stakeholders can be involved in changing and controlling their environments.

One of the challenges to those involved in participatory design has been to find ways of harnessing the creative potential of end-users, such as steelworkers or cash-desk operators, to solve problems related to their work, workplaces or work equipment. These people carry out the actual work tasks on a day-to-day basis, and have intimate knowledge of the problem areas. They often hold the key to the solution of design problems. The snag is that they are not generally expected to contribute to creative problem-solving, and therefore do not necessarily have the skills or experience to do so.

Group discussion methods such as focus groups have the potential to capture this latent problem-solving capability and also have many advantages when it comes to user acceptance of new tools or systems. They are often used in participatory design initiatives for these reasons. Over the years, those working in participatory design have developed and adapted focus groups and related techniques to meet these needs. This chapter reviews the ways such methods are used in this field and provides an overview of specific tools, including practical examples.

11.2 PARTICIPATORY DESIGN – A BRIEF OVERVIEW

This chapter is not about participatory design *per se* – for this, the reader is referred to Wilson and Haines, (1997), and Haines and Wilson (1998). To provide a broad understanding of participatory design and its philosophy, however, a brief overview is presented here, drawing mainly from research in human factors/ergonomics.

Participatory design as a concept has been a focus of attention within the human factors/ergonomics community since the early 1980s. Although some would say it is a relatively recent phenomenon, many would say that it has been around since the inception of ergonomics itself. It is not surprising that human factors/ergonomics and the concept of user participation should go hand in hand. User involvement in the design process is a central tenet to the discipline, and one of the things that makes human factors/ergonomics unique: indeed, it is often referred to as 'User-centred Design'.

11.2.1 Dimensions of Participation

Participatory design means different things to different people. Some would view the capturing of users' opinions with a questionnaire survey as an example of participation. Others would see this as a very diluted form, and would consider that fully empowered workers designing tools and systems is more representative. It is impossible, and probably undesirable, to create strict definitions and boundaries. It is helpful, however, to have a framework to describe the different dimensions of participatory approaches. An example is summarised in Table 11.1.

11.2.2 The Attractions of Participatory Methods

Although there is a wealth of data on human characteristics and numerous sources of design guidelines, it is hard to ensure that all key factors are taken into account and the correct compromises made during design development. Users are a key source of information and experience – they know more than anybody about the positive and negative features of the current design. If this knowledge can be harnessed to improve the overall design solution, then so much the better.

For projects with longer-term user involvement, for example, the design of work, workplaces or equipment, the benefits are potentially even greater. Participatory approaches can lead to greater ownership of the solutions, helping to make implementation easier (e.g. Imada and Robertson, 1987, and Wilson, 1995). Also, the experience of taking part can lead to benefits such as improved understanding and hence embedding and spreading of good design principles in the organisational culture (e.g. Daniellou and Garrigou, 1992, and Daniellou *et al.*, 1990).

11.2.3 Some Pitfalls and Problems

Whilst participatory approaches have many positive features, it is important to strike a note of caution. This is particularly important for projects carried out in the context of organisations, for example, workers participating in the design of work or workplaces. Such projects are rarely easy to set up and support, and may take more time and effort than traditional approaches (Zink, 1996). If not managed sensitively, the outcomes can actually be counter-productive. There can be an unwillingness to participate and suspicions that it will undermine existing roles, both management and trade union (Forrester, 1986). Also, if the programme is half-hearted or cynical (possibly even a manipulative programme with a covert agenda), then it will not provide long-term benefits. If the organisational climate is not conducive to participation, then it will not succeed. Further problems relate to the process of participation. Group based decision-making can be hard to manage, leading to the adoption of safe options or degenerating to the lowest common denominator (Fraser and Forrester, 1982). This emphasises the importance of adopting robust structures and methods.

Table 11.1 Dimensions of participation (adapted from Haines and Wilson, 1998)

Dimensions	Scales and categories		Notes
Extent/level	Organisation Work system Workplace Product	⇩ ⇩ ⇩	The scope of the participatory initiative. This can be wide, for example, across a whole organisation, or quite narrow – a specific product or tool perhaps.
Purpose	Work organisation Design Implementation	⇩ ⇩	At one end of the scale, participation may be a long-term feature of the way in which the organisation organises work. At the other end, it may be a specific initiative to implement a particular change.
Continuity	Continuous Discrete	⇩	Is the participation a continuous process or is it used just for one-off exercises?
Involvement	Direct (full/partial) Representative	⇩	With direct involvement, the stakeholders (or a sub-group of stakeholders) are directly involved in the process. Representative involvement is where representatives of the users (but not necessarily users themselves) take part in the process.
Formality	Formal Informal	⇩	Formal participation normally takes place through mechanisms such as teams or committees.
Requirement	Voluntary Compulsory	⇩	The best forms of participation are voluntary. In some cases, however, participation may be a compulsory job requirement, e.g. taking part in quality circles.
Decision-making	Workers decide Consensus Consultation	⇩ ⇩	The range extends from situations where workers decide for themselves, down to situations where managers make the decisions and workers are simply consulted.
Coupling	Direct Remote	⇩	The directness with which the outcomes of a participatory initiative are applied. People redesigning their own workplaces are examples of direct coupling. With remote coupling, there is some filtering, e.g. participants' views collected via a company questionnaire.

11.3 TOOLS AND TECHNIQUES FOR PARTICIPATORY DESIGN

11.3.1 Harnessing the Creativity of Non-experts

By its nature, participatory design requires the input of people who are not generally accustomed to being involved in problem analysis or the creation of solutions. Whether managers or workers, they are unlikely to have received much training in these skills, and are even more unlikely to have practised them. The problem is more acute if the area to be explored is complex, for example, the design of the user interface to a computer system.

One might assume that in order to have a meaningful input to such problems, it would be necessary to provide detailed training in specialist skills, for example, human factors/ergonomics, design or engineering. Indeed, some longer-term participatory initiatives have adopted this approach. For example, Kuorinka and Patry (1995) describe how a participatory group including workers were trained in motion-time-method analysis so that they could investigate how to improve the task of replacing fire-resistant vessel linings in a steel processing plant.

Many participatory projects can, however, be effective without the need to invest heavily in such preparatory training. The use of group discussion methodologies such as focus groups to facilitate non-expert input is a good example of this.

11.3.2 The Use of Group Discussion Methods in Participatory Design

The term 'group discussion' is used throughout this chapter as a broad description for methods that involve people working together in group activities to identify, analyse and solve problems. These methods may be predominantly discussion-based (for example, a basic focus group) or they may involve more structured processes incorporating hands-on activities. There are many different variations and adaptations, several examples of which appear in this book (see Chapters 2, 5, 8, 9 and 10).

Within participatory design, it is often helpful to provide a framework or structure to guide the participants through key phases of the design process, typically:

- identification of problems;
- analysis of problems to identify causes;
- identification of potential solutions;
- assessment of solutions;
- development/improvement of solutions.

Design Decision Groups or DDGs (Wilson, 1991) are an example of this approach. DDGs were derived from Shared Experience Events (O'Brien, 1981), and use a range of *thinking tools* within a pre-determined, but adaptable, structure or process (see Section 12.3.9 for an example). A further feature of DDGs is the incorporation of hands-on mock-up exercises that allow participants to experience their design ideas as they emerge.

DDGs have been used for many applications, including retail check-outs, library issue desks and control room designs. The basic format has seen several adaptations and extensions. One is the use of reference visits (Lehtela and Kukkonen, 1991) to enable participants to develop an awareness of solutions adopted in other situations, thus broadening their outlook. Another variation is the concept of Problem Solving Groups, where the participants' remit is extended to include activities such as 'costing up' the solutions as part of the evaluation process (Wilson, 1995).

11.3.3 Thinking Tools

Discussion group methods potentially are powerful ways of helping participants contribute directly to design solutions. To reap these benefits though, it is important to create an environment that not only elicits information from the participants, but also facilitates problem analysis and encourages creativity. Thinking tools are key to this.

Thinking tools have been used widely in participatory design projects. Many of these tools have been adopted from other fields and adapted as required. Others have been invented for specific purposes. Some of the thinking tools documented in the literature on participatory human factors/ergonomics are summarised in Table 11.2.

The techniques given here are just a sample: there are many more cases that have not been published, and new ways of doing things are being invented all the time. Even so, the examples given here provide a rich source of ideas and inspiration. As with virtually all techniques seen in this book, none of these tools has a very tightly defined methodology. All are extremely flexible and can be adapted to suit particular situations.

Table 11.2 Thinking tools used within group discussions for participatory design (adapted from Wilson and Haines, 1997). More detailed explanations for many of these methods are provided in Chapter 12 (continued overleaf)

Method	Purpose	Brief description	References
Cause-and-effect diagram	Problem analysis	Also called 'fishbone' or Ishikawa diagrams. Systematically looks at cause and effect, either to understand problems, or to identify solutions.	Imada (1991) (see 12.2.1)
Pareto Analysis	Problem analysis	Used to identify the main factors contributing to a problem and establish the relative impact, e.g. the frequency with which particular errors lead to the scrapping of parts.	Imada (1991)
Autoconfron-tation	Problem analysis	A video of a work process is shown to the group whilst the worker explains the work. Helps to reveal key problems and provides a starting point for identifying solutions.	Kuorinka (1997) (see 12.1.4)
Round-robin questionnaire	Creativity stimulation and idea generation	A series of open-ended statements which are completed by each participant in turn. Helps to identify problems and potential solutions.	O'Brien (1981); Wilson (1991) (see 12.1.9)
Word map	Creativity stimulation and idea generation	Participants volunteer words related to the topic, which are then recorded on a flip-chart. If required, relationships between words can also be shown.	O'Brien (1981); Wilson (1991) (see 12.1.11)
Drawing, or silent drawing	Creativity stimulation and idea generation	Participants use drawings to show their ideas. Drawings can be of any form and can be used as the basis for ongoing discussion or evaluation.	O'Brien (1981); Wilson (1991) (see 12.3.4)
Brain-storming	Idea generation and concept development	Used to quickly generate and capture ideas volunteered by participants. To encourage a wide range of ideas, discussion/criticism during the session is not allowed.	West (1994) (see 12.3.1)

Table 11.2 (continued)

Method	Purpose	Brief description	References
Scenario driven discussion	Idea generation and concept development	Scenarios of use for a product or system are used to help bring context to the session. Can help people imagine how they will use future products or systems.	Snow *et al.* (1996) (see 12.4.1 and Chapter 9)
Story-boarding	Concept development	Useful for designing change processes. Uses words and pictures to develop and assess different 'story lines', or approaches to the problem.	McNeese *et al.* (1995); Higgins (1994)
Layout modelling and mock-ups	Concept evaluation	Mock-ups, however crude, enable concepts to be visualised, and evaluated using simple simulations. 2D mock-ups can be used for 'layout' issues, e.g. equipment positions in a room. 3D mock-ups allow more in-depth assessment and fine-tuning.	Wilson (1991) (see 12.3.7 and 12.3.8)
Dynamic space manipulation	Idea generation and concept development	Similar to 2D layout mock-ups: useful for designing room or building layouts. A plan is projected onto a screen, and cut-outs of components re-positioned on this.	Snow *et al.* (1996)
Role playing and simulation games	Idea generation and concept development	Used in conjunction with layout modelling and mock-ups to test and develop ideas. Can be used in other ways, e.g. to explore complex processes and work flow.	Ruohomaki (1995); Wilson (1991) (see 12.4.5)
Virtual reality (VR) and virtual environments (VE)	Concept evaluation	Can be used instead of physical mock-ups to visualise concepts. Virtual tours help workplaces to be visualised, and allow different perspectives (e.g. a wheelchair user's) to be experienced.	Snow *et al.* (1996)
On-line illustrator	Idea generation and concept development	An illustrator captures verbal or sketch concepts and represents them visually to the group. Helps to clarify details, trigger new ideas and enable ideas to be evaluated.	O'Brien (1981) (see 12.3.5)
Product or system demon-strations	Problem analysis and concept evaluation	Provides background to enable participants to assess the product/system, and/or to incorporate it into their designs for future tasks, workstations or workplaces.	Wilson (1991) (see 12.4.2-12.4.4)
'Twinning' with user trials	Concept evaluation	Focus group discussion used to provide detailed feedback on the outcomes of a trial held in advance of the session.	Caplan (1990) (see 12.4.4)
Decision-making analysis	Concept evaluation	Analysis procedure within group discussion to determine relative importance of product features and to assess the extent to which different design concepts satisfy these.	Caplan (1990) (see 12.4.10)

11.4 CASE STUDIES

Many case studies related to participatory design have been published; for a detailed list, see Haines and Wilson (1998). The two examples given here, however, are used mainly to illustrate how group discussion type techniques have been used to analyse problems and create potential solutions. Two cases are given, one relating to an industrial workplace setting and the other to the design of an IT (information technology) system.

11.4.1 Case Study 1: Crane Control Room

Background

The setting for this example was the control room for the cranes used to feed refuse into an incinerator plant (for a detailed description, see Wilson, 1995). The project was prompted by health and safety concerns relating to the conditions in the control room, which were generally seen as unsatisfactory by the five drivers, their trade union and the site management. Problems included:

- poor working posture, including the need for drivers to hunch forward to view the bottom of the silos (see Figure 11.1);
- poor seating, and inadequate foot rests;
- poor control design and positioning;
- poor lighting, with reflections on the viewing window;
- isolation, poor communication and poor domestic facilities.

Figure 11.1. The drivers' position in the crane control room

The use of a participatory approach was appealing for several reasons: firstly, the solutions had to be acceptable to everyone concerned; secondly, the process had to be able to prioritise the key problems (due to technical and cost constraints); finally, the drivers did not want just a 'paper exercise', or to have 'findings that would be buried'. On the other hand, the drivers had not been asked to contribute to an analysis of their work before, and had limited formal education or problem-solving skills. So it was important that the procedure used was designed to enable them to contribute fully.

The approach

The overall approach is shown below. This was an adapted and extended version of the DDG (Design Decision Group) approach.

Stage 1: Familiarisation (for facilitators) with tasks, jobs, workplace and team.
Stage 2: Reference visits to similar sites.
Stage 3: DDG A – Drawing, discussion and idea generation.
Stage 4: DDG B – Layout building, discussion and idea testing.
Stage 5: Lighting and visibility simulation at site.
Stage 6: Sourcing and costing of solutions by participants.
Stage 7: Continual improvement after the change.

The purpose of the reference visits (Stage 2) was partly to see how similar problems had been handled elsewhere, but largely to stimulate interest amongst the drivers, and to get them thinking about their work and workplace in an analytical sense.

The first DDG session (Stage 3) was used to identify and prioritise the large number of problems, setting success criteria for any solutions and establishing a logical order for addressing problems. The techniques used during this session (and described earlier) were word maps, round-robin questionnaires and drawing. These activities were used to generate discussion; including critiquing problems and suggesting improvements.

Figure 11.2 Building and testing alternative control and seat set-ups in the DDG

The second DDG session (Stage 4) was used to focus on the design of the seats and control sets, which were the drivers' priorities. There were two main activities. Firstly, the drivers evaluated a wide variety of seats. This was not to choose one specific design, but to help them understand the range of possibilities when setting a specification for their final choice. The second key activity was the development and testing of different ideas for the control sets. The starting point for this was the ideas agreed in the first DDG session. The drivers used their metalworking and mechanical skills to test different types, angles and positions of controls, and to look at how these would fit with different seats (see Figure 11.2).

After seat and control design, the drivers' highest priority was the visual environment. During Stage 5, therefore, the drivers were given the opportunity to experiment with the temporary screens and blinds to minimise the problems with reflections in the control room. These quick and simple simulations enabled them to assess different arrangements cheaply and easily.

The purpose of Stage 6 was to define the preferred overall solution. Using the knowledge and experience gained from the earlier work, the drivers used catalogues (for example, control room seats and control sets) to select what they wanted to meet their needs and wants. An important feature of this exercise, however, was the need to take account of cost constraints that had been discussed and agreed with the site management team.

The outcome

The solutions resulting from the project led to a clear improvement in the crane control room, although some of the features could possibly have been improved further. More importantly though, when it came to implementation, there were no problems with the acceptance of the changes by the drivers, and both drivers and managers were very happy with the improvements. An encouraging later development was that interest and confidence had been raised such that a process of improvements to deal with some of the outstanding issues had continued (Stage 7).

11.4.2 Case Study 2: IT System to Support Safety Management

Background

In this example, a large distribution and delivery organisation were exploring the feasibility of implementing a company wide IT system to assist with the management of health and safety. The need was prompted by the increasing difficulties of managing paper-based systems for risk assessment records and accident reporting in over 1500 separate operational units. The key difficulties being encountered were:

- ensuring that all units had the latest information – and distribution costs;
- ensuring that tasks relating to compliance with regulations were being carried out;
- problems with the quality of information provided, e.g. accident investigations.

In the past, the company had experience difficulties in deploying safety management systems, in part due to lack of buy-in from key users. Up-front involvement was therefore important to establish the key user requirements. Also, it was critical to determine the acceptability in principle of an IT system to the largest section of users (local unit managers) who at the time were largely unused to new technology.

The approach

There were three main stages:

Stage 1: DDG sessions to establish key requirements.
Stage 2: Production of demonstrator system.
Stage 3: Focus group sessions to evaluate demonstrator system.

Stage 1 was to establish the key requirements for the system. This was done using a DDG approach to establish the good and bad features of the current systems and then to provide ideas for how future systems could work. There were three separate sessions, each held in a different part of the country to take account of the different needs for rural and urban locations. The sessions were approximately three hours long. The groups were made up of between six and ten people. The majority of these were local unit managers, but some area managers and safety professionals also took part. Also present were technical IT specialists who assisted with facilitation, but were present primarily to hear at first hand what was being said. The main steps in the Stage 1 sessions were as follows.

Introduction. This included explaining the project background and aims of the session, as well as personal introductions. 'Ground rules' for the session were also explained.

Round-robin questionnaires. Used as a 'warm-up', but also to establish the main good and bad points of current systems and to gain first impressions of the requirements for a new system (see Figure 11.3).

Discussion of the round-robin outputs. To ensure that everyone in the group understood what had been written and to tease out the key learning points.

Design exercise. The group was split into smaller groups of two or three people. The groups were asked to produce drawings and notes (on flip-chart pads) to illustrate their 'ideal' system, based on their own experiences and the key lessons from the preceding discussion. The groups were given some simple written instructions on the key things to consider and were encouraged not to be constrained by technical or financial considerations. Example outputs are shown in Figure 11.4.

Presentation and discussion. Each syndicate group presented their outputs, which were then critiqued and discussed by the whole group.

Conclusion. The main messages were summarised, the next steps for the project outlined and the group formally thanked for their contributions.

Of the current safety management tools and systems, the one I find most useful is...

Current accident report form
Safe systems of work manual
Risk assessment manual
COSHH register
Accident statistics
Accident book

The information/tools/support I would most like to have are...

A simple computerised package that the workforce understand
Something which is easily accessible and user friendly for all managers
Samples of all H&S aids
Time and simplified procedures
Local support

Figure 11.3 Example round-robin questionnaire outputs

Stage 1 clearly identified the key problem areas and the reasons for these. Many practical methods for solving the problems were identified. The use of computer-based systems was seen as being potentially very helpful and, in principle, acceptable to the users.

A key activity of Stage 2 was to analyse the Stage 1 findings and to produce a first version requirements specification. Using this, and some of the ideas produced in the DDGs, a demonstrator system was constructed. This system included all the features that were seen as key to a computer-based safety management system. It had sufficient functionality to enable navigation around the system but not for data entry.

Figure 11.4 Example drawings from design exercise in Stage 1

The objective of Stage 3 was to ensure that the key requirements of the users had been correctly interpreted. An additional aim was to gain reassurance that the use of a computer-based system was feasible in this operational context. A particular concern was that with up to 9000 potential users, it was important that the system was usable with minimal (or zero) training.

To do this, a series of focus groups were set up in the same locations as before. The groups were of similar size and approximately half the participants were the same. This provided useful continuity, but the introduction of some new blood gave the opportunity for a 'sense-check' on what was being proposed. The Stage 2 sessions had the following key activities.

Introduction. Background to the project and personal introductions.

Warm-up exercise. Syndicate groups were asked to brainstorm a list of all the safety management activities they had been involved with in the previous two weeks. This also provided further data on potential system requirements.

System evaluation exercise. After a brief introduction to the main features, the participants were given a list of tasks to complete using the system. This 'treasure hunt' provided the context for exploring the system, learning how to navigate through it and discover its key features along the way. As they were completing the tasks, participants were encouraged to note down any comments or difficulties they had. They were also observed as they worked on the tasks so that key problems and pitfalls could be identified. Note: participants were not provided with training or guidance, but had to carry out the tasks in a trial and error fashion.

Feedback questionnaire. On completion of the evaluation exercise, participants completed a questionnaire to provide feedback on system usability, ease of learning and usefulness to them in their day-to-day work.

Discussion and conclusion. The feedback questionnaire was used as a discussion guide to explore in detail the participants' views on the system and, in particular, to gauge its acceptability to them.

The outcome

The project provided a solid foundation for defining the proposed system's functionality. The DDGs used in Stage 1 were a quick and effective way of identifying the key problems with the current systems and identifying the kinds of tools that the users required. The inclusion of the evaluation/feedback loop (Stage 3) enabled the initial ideas to be refined and gave confidence that the system was feasible. In particular, it provided evidence that the goal of providing a system requiring minimal training was likely to be attainable.

The overall approach was well received by participants. Although the sessions were hard work, they found them stimulating and were pleased to have been able to contribute in such a practical way. The IT specialists taking part as observers/facilitators found the sessions extremely valuable because it helped them to understand the users and their needs. It also helped them to become a committed part of the project team.

11.4.3 Comparison of the Examples

It would be hard to find examples that are more different in terms of application and circumstances than those given here. For example:

- the crane example was in an industrial setting, with emphasis on the physical design aspects relating to comfort, health and safety. The IT example was concerned with a high-tech product involving predominantly administrative and mental tasks;
- the participants were quite different, with blue collar workers in one case and management grades in the other;
- in the crane example, there was a very small user group (five), whereas in the IT example the total user group was estimated as being up to 9000. Clearly, in the first case all the users could be involved in some way, whereas in the latter, only a small sample was possible. (Note: whilst small user groups have the advantage of permitting closer contact with the project, there can be problems if people do not wish to be involved or cannot be spared from the workplace to take part);
- with the crane control room, the majority of the project was within the control of the participatory group and engineers. With the IT system, development of the final product was inevitably going to require input and decisions from many different groups of people, including technical experts.

Although the cases were so different, the approaches used were similar in many ways; in particular the structured use of thinking tools. They provide examples of how group discussion methods can help to harness the analytical and problem solving skills of people, whatever industry they work in, or whatever types of jobs they do.

11.5 CONCLUSION

In the right circumstances and conditions, participatory design is a powerful and flexible way of solving or helping to solve a wide range of design problems. A key strength of the concept is the involvement of users and other stakeholders in the problem analysis and solution generation processes. The users have intimate knowledge of the problems with current tools, equipment or workplaces. The difficulty is in harnessing this unique knowledge because it is unusual for end-users to have the analytical and problem-solving skills usually associated with design.

Over the years a wide range of techniques have been developed and used to facilitate participatory design, particularly in the field of participatory human factors/ergonomics. Many of these techniques make use of group discussion methods such as focus groups and DDGs. Group discussion techniques are extremely flexible, and when linked with analysis and thinking tools to structure the process, help to harness the skills and knowledge of the participants.

As has been seen in many examples, including those described here, these methods will work with many different types of participants. They are also applicable to a wide range of problem areas including workplace design, job/task design, equipment/product design and IT system design.

It is clear that there is no single 'best method' for using these techniques. Each problem requires an approach that is tailored to suit the circumstances. There are a large number of thinking tools available to support these processes, a sample of which is described in this chapter and elsewhere in the book. There are undoubtedly many others that have never been reported in the literature, and new or modified techniques are being developed all the time.

11.6 REFERENCES

Caplan, S., 1990, Using focus group methodology for ergonomic design. *Ergonomics*, **33** (5), pp. 527-533.

Daniellou, F., Kerguelen, A., Garrigou, A. and Laville, A., 1990, Taking future activity into account and the design stage: participative design in the printing industry. In *Work Design in Practice*, Haslegrave, C.M., Wilson J.R. and Corlett, E.N. (eds), (London: Taylor and Francis).

Daniellou, F. and Garrigou, A., 1992, Human Factors in design: Sociotechnics or ergonomics? In *Design for Manufacturability and Process Planning*, Helander, M. and Nagamachi, M. (eds), (London: Taylor and Francis), pp. 55-63.

Fraser, C. and Forrester, K., 1982, Social groups, nonsense groups and group polarisation. In *The Social Dimension*, Tajfel, H. (ed.), (Cambridge: Cambridge University Press).

Forrester, K., 1986, Involving workers: Participatory ergonomics and the trade unions. In *Contemporary Ergonomics*, Oborne, D.J. (ed.), (London: Taylor and Francis).

Haines, H.M. and Wilson, J.R., 1998, *Development of a Framework for Participatory Ergonomics*, (Sudbury: HSE Books), ISBN 0 7176 15731.

Higgins, J.M., 1994, *101 Creative Problem Solving Techniques*, (Florida: New Management Publishing Company).

Imada, A.S. and Robertson, M.M., 1987, Cultural perspectives in participatory ergonomics. In *Proceedings of the Human Factors Society 31st Annual Meeting*, (Santa Monica, CA: Human Factors Society), pp. 1018-1022.

Imada, A.S, 1991, The rationale and tools of participatory ergonomics. In *Participatory Ergonomics*, Noro, K. and Imada, A. (eds), (London: Taylor and Francis), pp. 30-51.

Kuorinka, I. and Patry, L., 1995, Participation as a means of promoting occupational health. *International Journal of Industrial Ergonomics*, **15**, pp. 365-370.

Kuorinka, I., 1997, Tools and means of implementing participatory ergonomics. *International Journal of Industrial Ergonomics*, **19**, pp. 267-270.

Lehtela, J. and Kukkonen, R., 1991, Participation in the purchase of a telephone exchange – a case study. In *Designing for Everyone – Proceedings of the Eleventh Congress of the International Ergonomics Association*, Queinnec, Y. and Daniellou, F. (eds), (London: Taylor and Francis).

McNeese, M.D., Zaff, B.S., Citera, M., Brown, C.E. and Whitaker, R., 1995, AKADAM: Eliciting user knowledge to support participatory ergonomics. *International Journal of Industrial Ergonomics, Special Issue: 'Participatory Ergonomics'*, **15**, pp. 345-365.

O'Brien, D.D., 1981, Designing systems for new users. *Design Studies*, **2**, pp. 139-150.

Ruohomaki, V., 1995, A simulation game for the development of administrative work processes. In *The Simulation and Gaming Yearbook*, Saunders, D. (ed.) (London: Kogan Page), pp. 264-270.

Snow, M.P., Kies, J.K., Neale, D.C. and Williges, R.C., 1996, Participatory Design. *Ergonomics in Design*, **4** (2), pp. 18-24.

West, M., 1994, *Effective Teamwork*, (Leicester, UK: The British Psychological Society).

Wilson, J.R. and Haines, H.M., 1997, Participatory ergonomics. In *Handbook of Human Factors and Ergonomics* (2nd edition), Salvendy, G. (ed.), (New York: John Wiley and Sons), pp. 490-513.

Wilson, J.R., 1991, Design Decision Groups – A participative process for developing workplaces. In *Participatory Ergonomics*, Noro, K. and Imada, A. (eds), (London: Taylor and Francis). Also in Wilson, J.R., 1995, Ergonomics and participation, *Evaluation of Human Work*, Wilson, J.R. and Corlett, E.N. (eds), (London: Taylor and Francis), pp. 1071-1096.

Wilson, J.R., 1995, Solution ownership in participative work design: The case of a crane control room. *International Journal of Industrial Ergonomics*, **15**, pp. 329-344.

Zink, K.J., 1996, Continuous Improvement through Employee Participation. Some experiences from a long term study in Germany. In *Human Factors in Organizational Design and Management*, Brown, Jnr., V.O. and Hendrick, H.W. (eds), (Oxford: Elsevier Science), pp. 155-160.

PART III

Focus Group Tools

CHAPTER 12

Focus Group Tools

Joe Langford and Deana McDonagh

INTRODUCTION

As the contents of this book show, when ergonomists and designers use focus groups (or related techniques such as user group workshops or discussion groups), they adapt the basic focus group method to suit their specific requirements. Frequently, this involves introducing additional activities to the more normal discussion-based approach.

This part of the book summarises the main types of additional tools and activities that designers and ergonomists can employ to extend the effectiveness of focus groups. It is intended as a reference section, allowing the reader to identify methods that could be helpful to them and to provide sufficient information to get them started. These are useful both for those new to focus group methods and for those who are more experienced, but may be seeking inspiration for new ways of doing things.

The techniques and methods described within this section have been drawn from the contributing authors and from wider afield. The methods included are those considered most suitable for use within a focus group context and within the fields of human factors/ergonomics and design in particular. There are many more related tools and techniques described elsewhere, but these are not included because they are more suited to business and marketing applications. For more information on such methods, see Higgins (1994), Clegg and Birch (1999) and Greenbaum (1998).

Much effort has gone into finding examples of the tools and techniques used by ergonomists and designers to enhance focus groups. Those reported in human factors/ergonomics and design-related publications are often quite similar, even though they may be named differently and used for different purposes. Where there are obvious similarities and overlaps between methods, they have been grouped together in this book for the sake of simplicity.

USING THE METHODS IN THIS BOOK

There are many possible variations to the basic ideas described in the book. Indeed, whilst reviewing the literature to compile this summary, many examples of 'mixing and matching' of ideas were found. It is not intended that the descriptions given here should be strictly adhered to in a prescriptive manner. They are presented as basic techniques that can be adopted and adapted to suit specific purposes.

Note also that it is often helpful to use several different tools together. In Chapter 2, Anne Bruseberg and Deana McDonagh provide an example of a focus group plan that incorporates a range of different activities including visual evaluation of products, product handling and product personality profiles as well as conventional discussion sessions. In participatory design (see Chapter 11), the whole process can involve many different activities spread over several sessions, as in Design Decision Groups. It is common to find a whole range of activities linked together to enable the group to get to grips with the problem, generate solutions and then evaluate them, perhaps using mock-ups and simulations.

A word of warning: some of the techniques described here require participants to take part in activities that they might consider quite unusual. It is important that this is taken into account when planning discussion sessions. If participants are unfamiliar with a particular technique, this can result in them feeling uncomfortable or, worse still, lead them to resist or withdraw from the activity. It is extremely important that each participant feels comfortable and secure at all times.

HOW TO USE THIS CHAPTER

In this chapter, each technique is described briefly, along with the main advantages and disadvantages. In many cases examples are also provided. The descriptions of the methods are grouped by type.

Immersion and warm-up tools. Activities that are used to create an appropriate environment for effective discussion and idea generation to follow. These techniques are designed to prepare users in advance of the session, to get them in the mood for discussion, or simply to 'break the ice'.

Problem analysis tools. More formal, systematic tools that can be used to deconstruct a problem. These are useful as part of a session aiming to understand a particular problem before going on to find solutions, for example, in the context of a participatory design project. They might also be used by designers to help define a problem brief.

Idea generation and development tools. Techniques to enable the creation of new ideas and/or development of these. These often involve some form of hands-on activity. They may also involve, or be followed up by, some form of concept evaluation technique.

Concept evaluation tools. Activities that enable existing or proposed solutions to be assessed by participants. Some of these are relatively informal and unstructured, whereas others are more systematic, some of them involving numerical analysis. Concept evaluation is often the primary purpose of a focus group session, but is sometimes used as a starting point for the generation of new ideas.

Many of the techniques included here are multi-purpose. For example, brainstorming is an extremely versatile technique that could legitimately be placed in each of the above categories. In cases such as these, the primary or most appropriate use of the technique has been used to define its type.

The individual tools are listed in the following two index tables. In Table 12.1, they are listed by type, in the order in which they appear. In Table 12.2, they are listed alphabetically. Each method has a reference number to help locate it in the text.

Table 12.1 Tools and techniques listed by type and showing additional uses

Immersion and warm-up	Problem analysis	Idea generation and development	Concept evaluation		Ref. No.
				Tools primarily for immersion and warm-up	
•				Workbook	12.1.1
•				Diary/journal	12.1.2
•				Photographic record	12.1.3
•	•			Video record	12.1.4
•				'Day-in-the-life' exercise	12.1.5
•				Bring an object	12.1.6
•		•	•	Collage/mood board	12.1.7
•		•		Cognitive map	12.1.8
•				Round-robin questionnaire	12.1.9
•				Thought bubbles	12.1.10
•				Word map	12.1.11
				Tools primarily for problem analysis	
	•			Cause and effect analysis	12.2.1
	•			Why-why diagram	12.2.2
	•			Force field analysis	12.2.3
				Tools primarily for idea generation and development	
•	•	•	•	Brainstorming	12.3.1
		•	•	Nominal group technique	12.3.2
		•		How-how diagram	12.3.3
		•		Drawing	12.3.4
		•		On-line illustrator	12.3.5
		•	•	Storytelling	12.3.6
		•	•	Two-dimensional (layout) modelling	12.3.7
		•	•	Three-dimensional (form) modelling and mock-ups	12.3.8
	•	•	•	Design decision groups	12.3.9
				Tools primarily for concept evaluation	
		•	•	Scenario-based discussion	12.4.1
•			•	Visual evaluation of product/system	12.4.2
•			•	Product handling	12.4.3
•			•	Product/system user-testing	12.4.4
		•	•	Simulation and role playing	12.4.5
			•	Product personality profiling	12.4.6
•			•	Association	12.4.7
•			•	Conceptual mapping	12.4.8
			•	Attitudinal scaling	12.4.9
			•	Questionnaire	12.4.10
			•	Decision-making analysis	12.4.11
			•	Balance sheets or +/- charts	12.4.12
			•	SWOT analysis	12.4.13
			•	List reduction	12.4.14
			•	Voting and ranking	12.4.15

Table 12.2 Tools and techniques listed alphabetically

Immersion and warm-up	Problem analysis	Idea generation and development	Concept evaluation	Tool/technique	Ref. No.
•			•	Association	12.4.7
			•	Attitudinal scaling	12.4.9
			•	Balance sheets or +/- charts	12.4.12
•	•	•	•	Brainstorming	12.3.1
•				Bring an object	12.1.6
	•			Cause and effect analysis	12.2.1
•		•		Cognitive map	12.1.8
•		•	•	Collage/mood board	12.1.7
•			•	Conceptual mapping	12.4.8
•				'Day-in-the-life' exercise	12.1.5
			•	Decision-making analysis	12.4.11
	•	•	•	Design decision groups	12.3.9
•				Diary/journal	12.1.2
		•		Drawing	12.3.4
	•			Force field analysis	12.2.3
		•		How-how diagram	12.3.3
			•	List reduction	12.4.14
		•	•	Nominal group technique	12.3.2
		•		On-line illustrator	12.3.5
•				Photographic record	12.1.3
•			•	Product handling	12.4.3
			•	Product personality profiling	12.4.6
•			•	Product/system user-testing	12.4.4
			•	Questionnaire	12.4.10
•				Round-robin questionnaire	12.1.9
		•	•	Scenario-based discussion	12.4.1
		•	•	Simulation and role playing	12.4.5
		•	•	Storytelling	12.3.6
			•	SWOT analysis	12.4.13
•				Thought bubbles	12.1.10
		•	•	Three-dimensional (form) modelling and mock-ups	12.3.8
		•	•	Two-dimensional (layout) modelling	12.3.7
•	•			Video record	12.1.4
•			•	Visual evaluation of product/system	12.4.2
			•	Voting and ranking	12.4.15
	•			Why-why diagram	12.2.2
•				Word map	12.1.11
•				Workbook	12.1.1

12.1 TOOLS FOR IMMERSION AND WARM-UP

The main objective of these tools is to prepare the ground for the discussion session and, in particular, to help ensure that the participants are ready and able to contribute fully. This can be achieved in the following ways:

By *preparing* participants prior to the session, for example, by giving them some pre-session work to do so that they have a heightened awareness of the areas to be discussed, and bring with them relevant experiences from daily living.

By *breaking the ice*, or providing warm-up activities at the start of a session so that the participants become used to the idea of taking part right from the outset.

By getting the participants fully *tuned in* to the subject in question and actively thinking about the issues at the start of the session, prior to more detailed discussion or other activities.

There are no hard and fast rules for which of these tools should be used in a given application. The choice is highly dependent on the topic of interest and/or the participant types. Pre-meeting or 'homework' activities are good for making the best use of time in a discussion session, but if there is a high risk of participants not completing them, it might be better to plan a warm-up exercise for the start of a session instead.

12.1.1 Workbook

This is an example of a pre-meeting immersion tool, as described by Liz Sanders and Colin William (Chapter 10). The workbook is sent to the participants for completion prior to the discussion session. The act of completing the workbook helps the participants to take note of their experiences and thoughts about the topic of interest. It gives them time to *tune in* before the session.

The format and content of the workbook are dependent on the purpose of the session. Typically though, the book will contain different types of questions such as demographic details, opinions and information about things people own and use. This enables the researchers to find out more about participants, but without using up valuable time in the session.

The workbook can also include hands-on exercises, for example, keeping a diary or journal (see Section 12.1.2), or making a photographic record (see Section 12.1.3). These enable the participants to think about the issues to be discussed and about their own experiences in advance of the session. If exercises are included, it is important to provide sufficient workspace and clear instructions. Also, the exercises need to be as interesting as possible to increase the chance of people completing them.

The completed book can be brought to the session, or mailed back in advance. If sent back in advance, the researchers have more time to use the information to help plan the session, for example, develop more relevant follow up questions. The workbooks can be used within the session as the basis of an activity to stimulate discussion, for example, asking participants to present the answers they have given to some of the exercises. Alternatively, individual participants can use their books as a memory aid during the session.

Main advantages:

- participants have the opportunity to tune in to the topic in advance of the session;
- workbooks can be used to stimulate discussion or act as a memory aid;

- makes effective use of time in the discussion session. Information not suited for discussion can be captured outside of the discussion time;
- a considerable amount of data can be captured for subsequent analysis if required.

Main disadvantages:

- time required by researchers to prepare and distribute the workbooks;
- relies on participants completing the workbooks. If they are not completed, the effectiveness of the session may be reduced.

12.1.2 Diary/journal

Diaries and journals can be valuable pre-meeting immersion activities. They may be included as part of a workbook exercise (see Section 12.1.1) and possibly be linked with making a photographic record too (see Section 12.1.3). Keeping a record of their actual activities in a given day, whilst carrying out certain tasks or using a product or system helps people to become immersed in the experience. It can also attune them to details of those activities that they normally take for granted.

In Chapter 8, Peter Coughlan and Aaron Sklar describe a project on mini-van interiors where they asked participants to keep a driving log for three days prior to the session. This included notes of trips they went on, who rode with them, the destination, how well the vehicle served their needs and their emotional state during each trip. The participants brought the journals to the discussion session where the notes and records served as prompts to their earlier experiences.

If sent back in advance, the information in participants' diaries/journals can be used to help plan the session and prepare questions. They can also be used in activities within the session, for example, asking participants to describe their actual experiences and emotions on a given day. The diaries/journals can also be used as a memory aid for participants during the session.

Main advantages:

- participants have the opportunity to tune in to the topic in advance of the session and are prepared for the discussion;
- diary/journal notes can be used to stimulate discussion or act as a memory aid;
- a great deal of information can be captured for subsequent analysis if required.

Main disadvantages:

- relies on participants completing the diaries or journals. If they are not completed, the effectiveness of the session may be reduced;
- some participants might not find writing easy or be comfortable writing;
- some participants may leave completion of the diary/journal until just before the session, thus limiting effectiveness and accuracy.

12.1.3 Photographic Record

The use of photographs adds a level of richness to purely text-based methods for recording information. A picture of a workspace can give deeper insights into the ways that people organise themselves, the ways they work and the things they use. Workbooks (see Section 12.1.1) and diaries/journals (see Section 12.1.2) can be greatly enhanced by giving people cameras to document their experiences. The resulting pictures can be mounted in the workbook or diary along with their notes.

If people are given cameras in advance of the session, they can take pictures to record snapshots of their experiences, objects and surroundings related to the topic of interest. The pictures help people to remember things once away from the actual setting. Or they can be used to prompt discussion during the session. Images brought in by one participant can help others to understand and respond to their points of view. If sent back to the researcher in advance, the pictures provided by the participants can also be used to help in the planning of a discussion session, for example, formulating questions.

There are some practical issues concerned with the use of cameras. Each participant will probably need to be provided with a camera. Disposable cameras are a relatively inexpensive option, and they are small and neat to carry. Time is required for film processing however. Polaroid cameras and film are much more expensive, but they allow people to take and evaluate pictures on the spot. They can also annotate them immediately if required.

Main advantages:

- participants have the opportunity to tune in to the topic in advance of the session;
- relatively quick and easy thing for people to do;
- pictures can be used to stimulate discussion or act as a memory aid.

Main disadvantages:

- relies on participants taking the pictures. If participants fail to do this, it may limit the effectiveness of the session;
- cameras have to be provided to some or all of the participants. This adds to the expense and organisational costs;
- time is required for film processing if disposable cameras are used.

12.1.4 Video Record

This technique uses a video recording as a discussion stimulus, for example, to explore work processes. The video recording is shown to the whole group whilst the person using the system describes what is going on, including any comments on problems and good features. The exercise focuses participants on the issues in question, and provides a useful stimulus to subsequent discussion. It can also serve as the basis for more detailed analysis of the problems and subsequent identification of potential solutions.

This method has been used to analyse workplace problems (Kourinka, 1997, refers to the technique as *autoconfrontation*) and clearly has potential for use with a wide range of topic areas. These include product usability and human computer interfaces. Video recordings of user trials, for example, can be used to stimulate discussion during a group feedback session.

The main drawback with the method is the time required to make and edit/compile the video recordings. This would generally need to be done by the researcher to ensure quality and consistency. Some topic areas may be hard to capture on video, for example, experiences whilst travelling.

Main advantages:

- provides strong focus and stimulus for discussion and probing;
- brings real-world context into the session.

Main disadvantages:

- time and effort required by researchers to make video recordings;
- may be inappropriate or impractical for some topic areas.

12.1.5 'Day-in-the-Life' Exercise

In such exercises, people outline their typical day, either for specific aspects (for example, use of computers, travelling, work tasks) or more generally. In cases where pre-meeting immersion tools such as diaries or journals (see Section 12.1.2) are impractical, a 'day-in-the-life' exercise is a good way of stimulating participants at the start of a session. Ede (1999) used the technique to learn more about the *real* activities of systems administrators in focus groups set up to study work practice.

It is helpful if people are given the opportunity to think quietly by themselves and make notes prior to presenting their typical day to the other participants. It is sometimes possible to ask people to produce simple *time-line* diagrams as shown in Figure 12.1. In her study, Ede co-opted other group members to write down on note cards each different activity as it was mentioned. These were then sorted and discussed later on in the session.

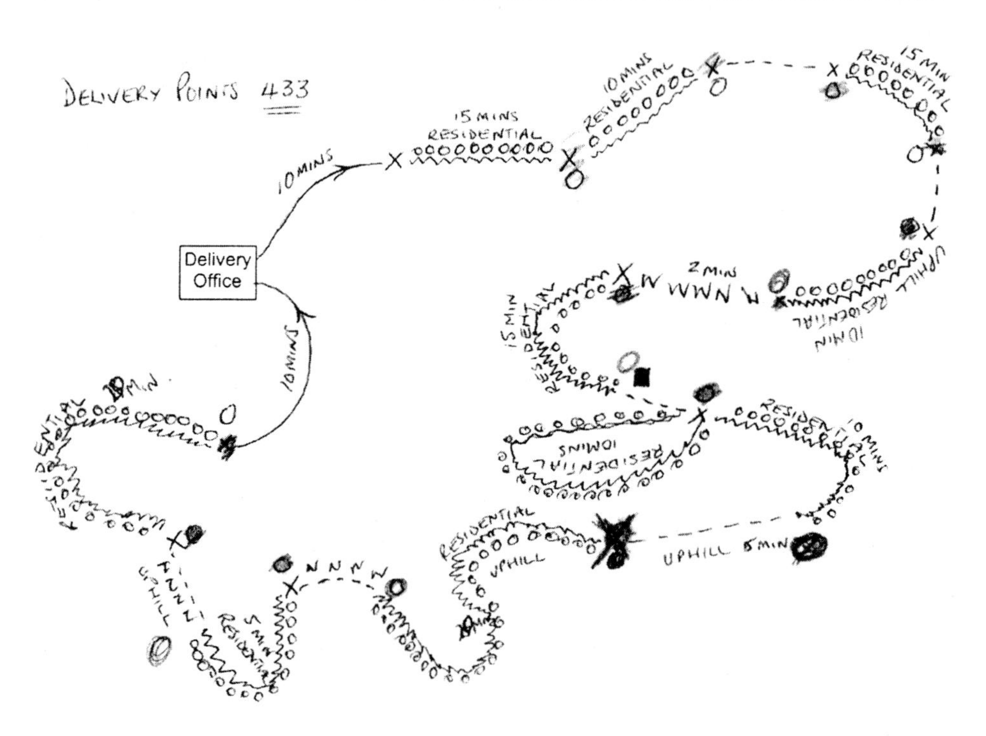

Figure 12.1 Time-line diagram for a mail delivery (courtesy of Royal Mail)

By carrying out the exercise, individuals are required to focus on the area of interest, reminding them of some of their real world experiences, which can then help stimulate the discussion. The time-line of events also provides a framework for people to explain their experiences and the emotions attached to them. 'Day-in-the-life' exercises act as valuable warm-up exercises to help people to become at ease with the discussion session environment and to get them involved.

Main advantages:

- no pre-work required from participants or researchers;
- quick and easy exercise for people to do;
- valuable warm-up exercise to help make people at ease and to get them involved;
- provides a stimulus for discussion and probing.

Main disadvantages:

- the descriptions may not be accurate;
- participants may wish to keep some details private;
- participants may disregard and omit details that they think are unimportant, but which may be critical.

12.1.6 Bring an Object

By bringing an object along with them to a focus group session, participants are encouraged to reveal some of their preferences, discuss their personal possessions or represent their everyday behaviours. Peter Coughlan and Aaron Sklar provide examples in Chapter 8.

In the first, a session facilitated for a toy manufacturer, children and parents were asked to bring their favourite toys. The discussion began with everyone explaining why they brought that particular toy. As the participants showed the toys to the group, the researchers were able to learn more about the roles of toys in their lives and what they valued about them. In the second example, a project about handheld devices, participants were asked to bring in an object they thought was pleasant to hold. This allowed the researchers to learn what values would apply when the participants assessed the handheld prototypes they would be shown later in the session.

As with photographic records (see Section 12.1.3), using objects as a focus for discussion provides an excellent mechanism for allowing researchers to learn more about the participants' behaviours, emotions and opinions. Such exercises also help people to become at ease with the discussion session environment and to get them involved.

Main advantages:

- minimal pre-work required from participants;
- good warm-up exercise to help make people at ease and to get them involved;
- provides a focus for gaining deeper insights into individuals' emotions and opinions.

Main disadvantages:

- relies on participants bringing objects. The effectiveness of the session may be limited if this is not done;
- some individuals might feel exposed or feel that they will reveal too much of themselves by bringing an object from home.

12.1.7 Collages and Mood Boards

Although there may be subtle differences between collages and mood boards, their use in the context of group discussions is similar. For the purposes of this section, therefore, they are treated as the same thing.

The main function of collages and mood boards is to enable people to articulate their experiences and feelings through pictures and words. They transcend linguistic limitations, overcoming the relatively narrow scope of words. For some individuals, it is easier to communicate their thoughts using pictures, and they may be more willing or able to put together a collage than write things down. Such techniques work well with children for these reasons. Other benefits of collages and mood boards include:

- the act of putting together a collage or mood board helps activate feelings and memories;
- people can use them to explore and express their dreams and aspirations;
- the completed collages and mood boards provide valuable stimuli for discussion amongst the participants, and probing by the moderator (see also association, Section 12.4.7);
- when the session is over, the outputs can be used to help present findings to clients or to provide designers with sources of inspiration.

Figure 12.2 Children assembling collages on *play* (from McDonagh *et al.*, 2002)

McDonagh *et al.* (2002) used mood boards in their study to explore the essence of *play* with regard to the design of playground equipment. In participative workshops, children were given a diverse set of images. They were asked to select those that represented play and assemble them on a sheet of paper (see Figure 12.2). They were also asked to provide a brief explanation of their creations, and the reasons for choosing each image. The results showed a high degree of symbolism, and particular 'triggers' from their personal experiences and memories were expressed. For example, play was expressed by many of the children as a social activity.

The basic requirements for a collage or mood board exercise are simple – all that is needed is a set of pictures and words (typically between 50 and 100), and a space on which to stick them down. In some cases, it can be helpful to provide samples of materials and textures too. A simple way of providing materials for the participants is to print the images and words on sheets and provide scissors and glue for the participants to cut out and stick down for themselves. It is much easier and quicker for participants though if they can be given pre-prepared sticker sheets. In Chapter 10, Liz Sanders and Colin William describe how to create a collage kit.

Main advantages:

- transcends limitations of words by using images as primary method for articulating thoughts. Suitable for use with children;
- activates participants' feelings and memories during production and enables them to express dreams and aspirations;
- provides a stimulus for discussion and probing;
- outputs can be used to help present findings and to provide inspiration for designers.

Main disadvantages:

- requires thorough preparation;
- requires reasonable amount of time in the session to create the collages (typically a minimum of 15 minutes);
- unfamiliar exercise that may be uncomfortable for some participants. Clear briefing and some assistance from the moderator may also be required.

12.1.8 Cognitive Maps

Cognitive mapping provides insights into the underlying cognitive structures that people have. It enables better understanding of how individuals think about complex phenomena. When they create a cognitive map, participants plot out their understanding of systems and how things fit together or relate to each other. Symbols, shapes and sometimes words are used to show connections, clusters or hierarchies of concepts. There are no conventions or rules – individuals use the representations that work best for them.

As with collages or mood boards (see Section 12.1.7), the method is predominantly visual in nature, and encourages participants to express thoughts and ideas without being limited by language. When they have made their maps, the participants are asked to explain them. This provides opportunities for others to join the discussion and the moderator to probe further. The technique can be used to help individuals describe current scenarios. For example:

- how a process or system works (or how the participant thinks it works);
- how a person understands an event;
- how a person makes decisions about an issue.

It can, however, also be used as a creative tool, allowing participants to explore and express ideas for how they would like things to work in the future. Completed maps provide helpful stimuli for group discussion and the exploration of concepts. They also provide useful aids for presenting outputs to stakeholders such as clients, or inspiration for designers.

Creation of a cognitive map requires coloured pens, a set of symbols and words, and somewhere to stick them down (see Figure 12.3). As with collages or mood boards, the symbols and words can be pre-printed on sheets for the participants to cut out and stick down for themselves, or provided on sticker sheets. Many pre-prepared symbols and shapes are readily available from art and craft stores. In Chapter 10, Liz Sanders and Colin William describe how to create a cognitive mapping kit.

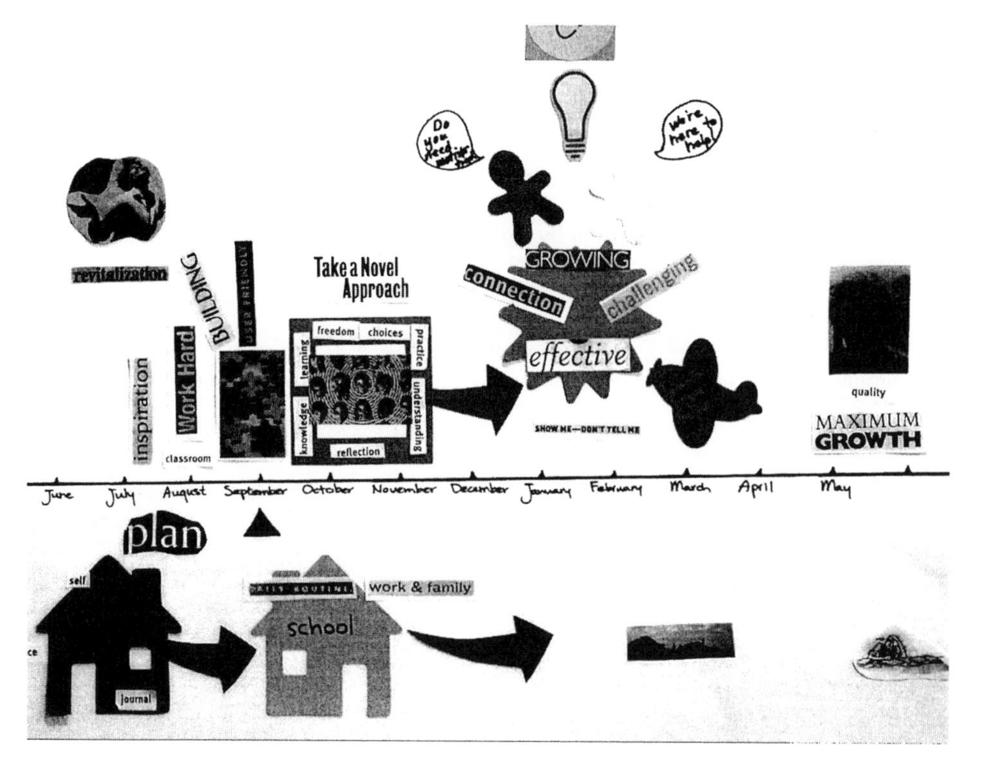

Figure 12.3 Cognitive map completed by a school teacher expressing thoughts and ideas about her ideal teaching experience for the upcoming school year (courtesy of SonicRim)

Main advantages:

- transcends limitations of words by using shapes and spatial layouts to help articulate thoughts, and aid understanding of underlying cognitive structures;
- provides a stimulus for discussion and probing;
- provides a vehicle for generating design ideas;
- outputs can be used to help present findings and to provide inspiration for designers when creating new concepts.

Main disadvantages:

- requires thorough preparation;
- requires time in the session to create the maps (typically a minimum of 15 minutes);
- unfamiliar exercise that may be uncomfortable for some participants. Clear briefing and some assistance from the moderator may also be required.

12.1.9 Round-robin Questionnaire

The round-robin questionnaire is a relatively simple technique for getting participants involved in the session and focused on key issues right from the start. In some circumstances, it also enables problem areas to be analysed, and potential solutions to be identified.

The technique involves a series of blank sheets of paper with an open-ended statement on the top. The statements used must be tailored to the topic of the discussion. Some examples are shown below:

The hardest part of my job is...
The design change that would most improve my workstation is...
The best piece of equipment I use in my job is...because...
The worst piece of equipment I use in my job is...because...
The thing I like best about working in my company is...
The thing I like least about working in my company is...

Ideally there should be as many sheets/statements as there are participants, but a small mismatch is acceptable. One sheet is issued to each participant, and all spend a minute or two completing the statement in the way they think most appropriate. When everyone is ready, the sheets are passed on to the person sitting to one side, so that everyone has a new statement to complete, following on from the answers given previously. This is repeated until all participants have completed all the statements (see Figure 12.4), typically ten to fifteen minutes.

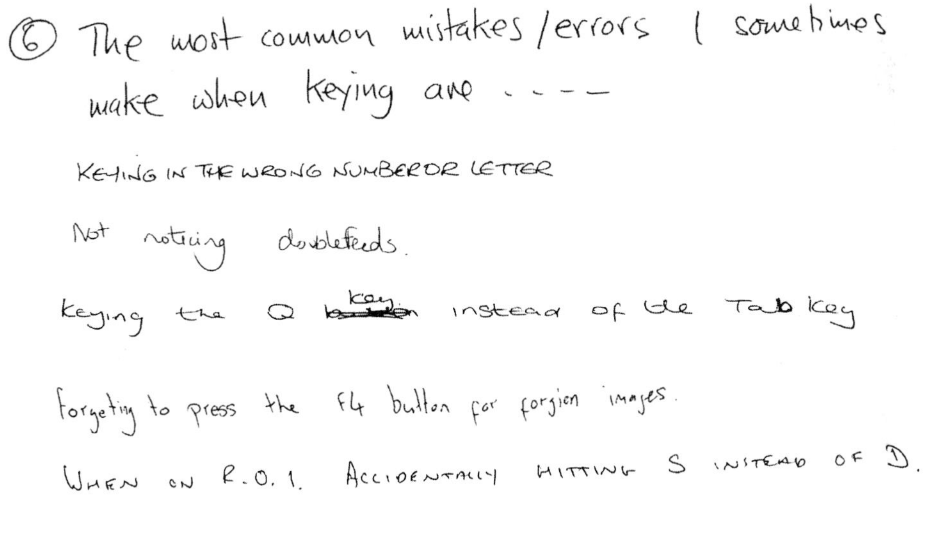

(6) The most common mistakes/errors I sometimes make when keying are ----

KEYING IN THE WRONG NUMBER OR LETTER

Not noticing doublefeeds.

keying the Q ~~button~~ key instead of the Tab key

forgeting to press the F4 button for forgien images.

WHEN ON R.O.1. ACCIDENTALLY HITTING S INSTEAD OF D.

Figure 12.4 A completed round-robin questionnaire from a focus group held to assess a data entry task

Whilst completing the sheets, individuals can draw inspiration from the answers that others have given previously if they wish. Sometimes a person will find that someone else has already given the answer they would have chosen, in which case they should give their second (or third) choice answers to avoid repetition. In this way, when the exercise is complete, a wide range of issues will have been identified.

The completed sheets have two main uses. Firstly, they provide a written record for later analysis if required. Secondly, and more importantly, they can be used as the basis for a discussion session, enabling participants to elaborate on the answers given. For this reason, it is sometimes useful to use flip-chart size paper for the sheets, with answers written using bold markers so that everyone can easily view the sheets when they are attached to a wall or easel.

Main advantages:

- no pre-work required from participants;
- minimal preparation for researchers;
- valuable warm-up exercise to help participants relax and to get them involved;
- provides stimulus for discussion and probing;
- provides written record for subsequent analysis if required.

Main disadvantages:

- consumes time during the session;
- some participants may feel they are being ‘pressured’ to contribute;
- some participants may find the exercise difficult to complete.

12.1.10 Thought Bubbles

This is a quick and simple way of engaging the participants in the topic of interest at the start of a session. The technique works by getting individuals to place themselves in the shoes of a person in a given situation related to the topic of interest. A simple drawing or image of the scenario is given to, or shown to, the participants and they are asked to write down what might be going through the mind of the person involved (see Figure 12.5).

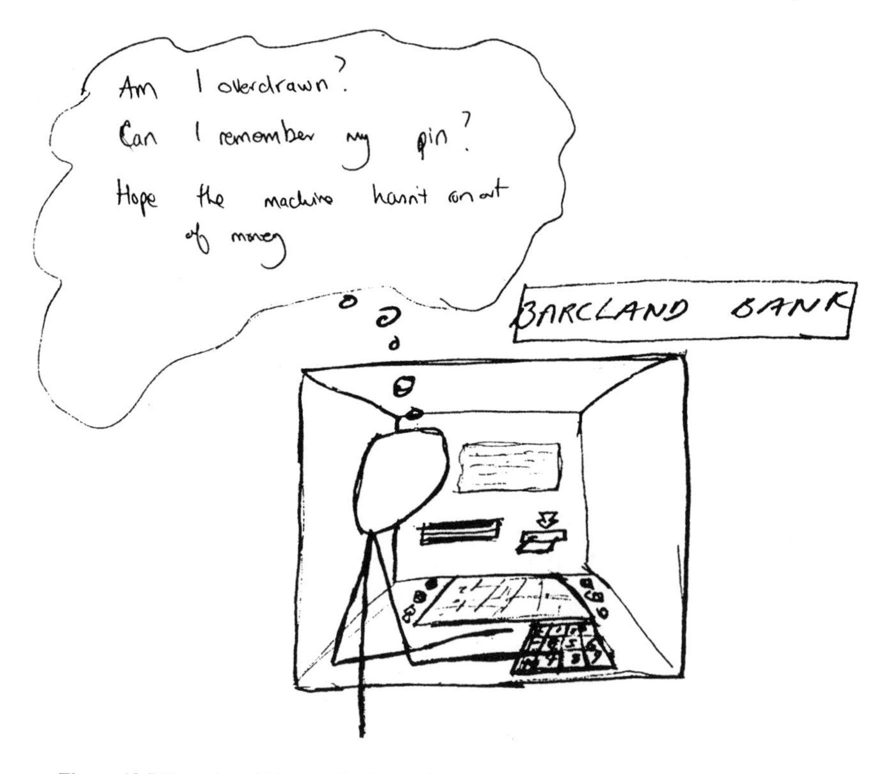

Figure 12.5 Thought bubble exercise from a focus group on alarm panel design (courtesy of Royal Mail)

In this example, the focus group workshop was held to understand the sorts of problems individuals have when setting, unsetting and programming intruder alarms.

Scenarios shown in the thought bubble diagrams in this case included related scenarios (using a video recorder and a bank cash point or ATM) as well as the actual scenario in question.

The exercise is relatively quick to set up and can be run at the start of a session to help participants to focus on the topic of interest. The outputs can be used to stimulate discussion and can also provide a source of material for subsequent analysis or when reporting results.

Main advantages:

- no pre-work required from participants;
- relatively uncomplicated exercise to do;
- useful warm-up exercise to help participants relax and to get them involved;
- provides stimulus for discussion and probing.

Main disadvantages:

- consumes time during the session;
- participants may be unfamiliar with the approach;
- the variety of responses can be limited.

12.1.11 Word Map

Used by O'Brien (1981) and Wilson (1991), a word map is a tool that helps participants focus on the topic of interest at the start of a session, and to understand the main linkages between them. It is similar to *mind mapping* (Buzan and Buzan, 1993) but adapted for use in group sessions.

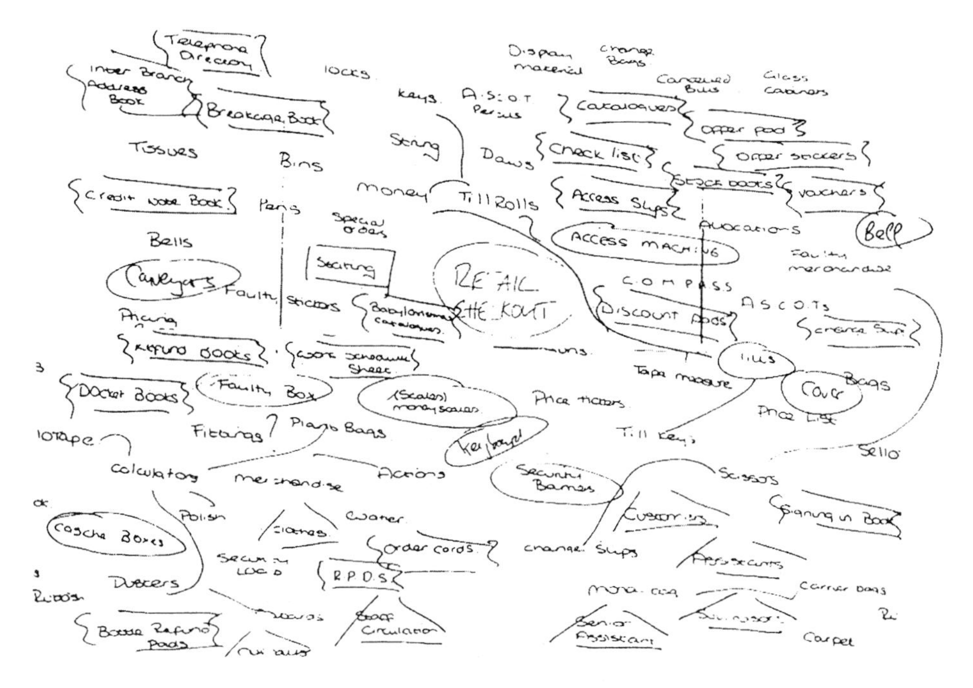

Figure 12.6 Part of a simple word map (from Wilson, 1991)

With O'Brien's technique, each person in turn is asked by the moderator to volunteer a single word relating to the topic in question. These can be written directly onto a flip-chart, or written on cards, or post-it notes and attached to a wall or display board. In a second round, the participants are asked to provide secondary words, linked in some way to a word that has already been put up. These are also attached to the wall, alongside the primary word, or linked with a line. The process continues until the supply of relevant words is exhausted. The end result is a word map showing key areas of the topic and their relationships (Figure 12.6).

As a variation, participants can volunteer the words at random, with no attempt to record any linkages. This is similar in many ways to a brainstorming session (see Section 2.3.1).

The exercise is relatively quick to set up and run. If done at the start of a session, the outputs can be used to stimulate discussion or, in a more formal way, to create an agenda. They can also be used as a source of information for subsequent analysis if required.

Main advantages:

- no pre-work required from participants;
- creates a common understanding of issues and their relationships;
- useful warm-up exercise to help participants relax and to get them involved.

Main disadvantages:

- consumes time during the session;
- some participants may feel 'pressured' to contribute;
- requires a skilled moderator to help organise and manage the inputs.

12.2 PROBLEM ANALYSIS

The techniques described here are relatively uncomplicated analytical tools that can be used in the context of a group discussion. They can be used to help understand a problem and deconstruct it so that the key elements and the linkages between them can be better understood. Several of the techniques described here are not specific to ergonomics or design, and have been adapted from other, more general tools used in management.

Such techniques are unlikely to be employed in discussion sessions aimed at learning about what participants think and feel about consumer products or services. They are, however, well suited to participatory design problems where users, operators or other stakeholders are solving problems concerning workplaces, systems or equipment for their own use.

12.2.1 Cause and Effect Analysis

This is a systematic way of looking at cause and effect, either for understanding the causes of a problem, or for identifying what needs to be carried out to achieve a desired effect. The analysis is carried out using cause and effect diagrams, which are sometimes called 'fishbones' or 'Ishikawa diagrams' after their inventor (see Higgins, 1994). The basic procedure for carrying out cause and effect analysis is as follows.

Carry out a brainstorming session (see Section 12.3.1) to identify the causes that might lead to the effect under analysis.

Group the results of the brainstorming session under common main headings or themes. With the lamp example (Figure 12.7), 'power failure', 'no house current' and 'wall switch turned off' are all causes linked to 'power'. Organise causes under further sub-groupings where required, for example, 'power failure' may be caused by a storm or power plant failure.

Create the fishbone diagram as shown in Figure 12.7 using the side 'bones' for main headings and marking the main causes and sub-causes alongside these.

Review the diagram adding further causes or re-organising them as required.

Prioritise causes identifying the most important for further investigation or action as required.

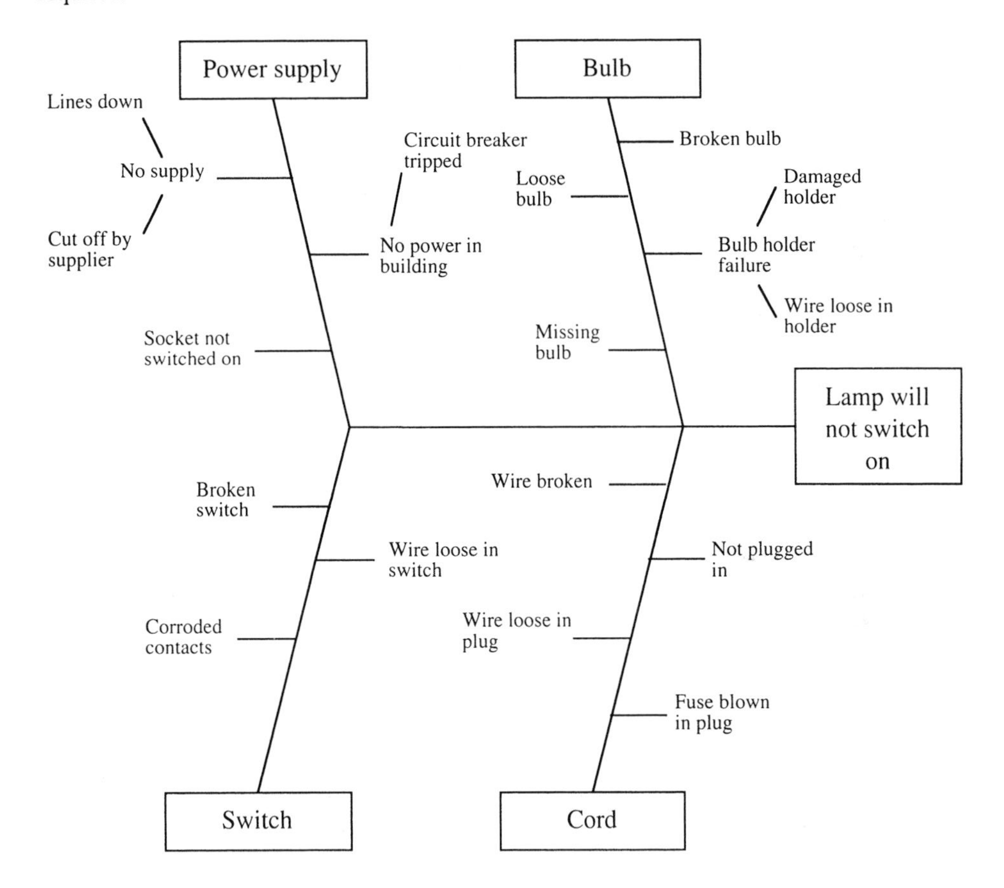

Figure 12.7 Example fishbone diagram to analyse why a lamp does not work

No analysis should take place in the early stages (the brainstorming phase). During this phase, participants should be encouraged to mention as many ideas as possible. It is only after the initial idea generation that analysis should take place, i.e. testing the understanding of the ideas, grouping, combination and prioritisation. There is no cut-off point for new ideas to be added; it is likely that the construction of the diagram will trigger new thoughts in itself.

It is important that all participants can see the diagram when it is constructed, so large charts and large writing are likely to be required. It may be helpful to transfer the causes onto cards or post-it notes that can be placed onto a wall display. For complex problems, diagrams can become unwieldy to work with. In these cases, one or more of the side bones can be removed and dealt with separately as fishbones in their own right.

The technique is not limited to the analysis of existing problems. For the design of a new system, for example, it can be used to help understand the key things that need to be in place to ensure a satisfactory outcome. In other words, the *effect* is the desired end-state and the *causes* are the actions or functions needed to achieve this. The potential application areas are wide ranging, including work systems design, organisation design, marketing strategies, project planning and identification of system requirements.

Main advantages:

- uncomplicated but powerful analytical technique to identify key issues affecting a problem and to understand the linkages between these;
- harnesses the collective knowledge and skills of the participants;
- not limited to the analysis of existing problems – can be used to identify ways of achieving the desired end-state in wide range of application areas.

Main disadvantages:

- requires a skilled moderator to help organise the inputs;
- participants may be unfamiliar with the method;
- can be time-consuming.

12.2.2 Why-why Diagram

Problem	WHY?	WHY?
Lamp does not work	No power to building	Lines down
		Cut off by supplier
	No power to lamp	Lamp not plugged in
		Socket not switched on
		Circuit breaker tripped
	Bulb failure	Missing bulb
		Broken bulb
		Loose bulb
	Failure in power cord	Fuse blown in plug
		Loose wire in plug
		Wire broken
	Failure in bulb holder	Damaged bulb holder
		Loose wire in holder
	Failure in switch	Broken switch
		Corroded contacts
		Loose wire in switch

Figure 12.8 Example why-why diagram

This is a variation on cause and effect analysis (see Section 12.2.1). Why-why diagrams explore the reasons why a particular problem may be occurring by asking the question 'why?' and progressively re-describing causes at greater levels of detail (see Figure 12.8).

A variation of this method can be used in reverse to explore ways of solving problems. This is the how-how diagram described in Section 12.3.3.

Main advantages:

- uncomplicated but powerful analytical technique to identify key issues affecting a problem and to understand the linkages between these issues;
- harnesses the collective knowledge and skills of the participants.

Main disadvantages:

- requires a skilled moderator to help organise the inputs;
- participants may be unfamiliar with the method;
- can be time-consuming.

12.2.3 Force Field Analysis

Force field analysis helps identify the factors that help or hinder the achievement of a task or other desirable outcome, for example, achieving production or quality targets, or implementing a new system. Identification and subsequent manipulation of these factors can help close the gap between the current state and the desired end-state. This can be a useful technique for participatory design projects.

The basic procedure for carrying out a force field analysis is as follows.

Create the chart. As shown in Figure 12.9, draw vertical lines on a flip-chart, one down the centre (to represent current state) and one to the right (desired end-state).

Identify *helpers* and *hinderers*. Use a brainstorming session (see Section 12.3.1) to list the helping and hindering factors or *forces* and write these on the sheet – helpers to the left and hinderers to the right.

Estimate the relative strengths of the forces. Mark them on the flip-chart sheet either as scores (e.g. 1 is weak, 5 is strong) or as arrows with different lengths (e.g, short arrow is weak, long arrow is strong).

When the analysis is complete, the group can then use the information to generate potential actions by considering how to increase the number and strength of helping forces, and vice versa for the hindering ones.

The technique helps key issues to be identified, and prioritised for action. It is a valuable tool to aid planning, and as such has many potential uses within participatory design in the workplace. It also has potential application outside of this in understanding the views of consumers regarding the factors affecting the successful introduction, use or uptake of a new service. It is not suited to the exploration of design details relating to products or systems.

Main advantages:

- uncomplicated but powerful analytical technique to identify and estimate the strength of key factors affecting achievement of a desired outcome;
- can be used both for planning and for post-implementation analysis purposes.

Main disadvantages:

- limited use for consumer products or services;
- not suited to exploration of design details for products or systems.

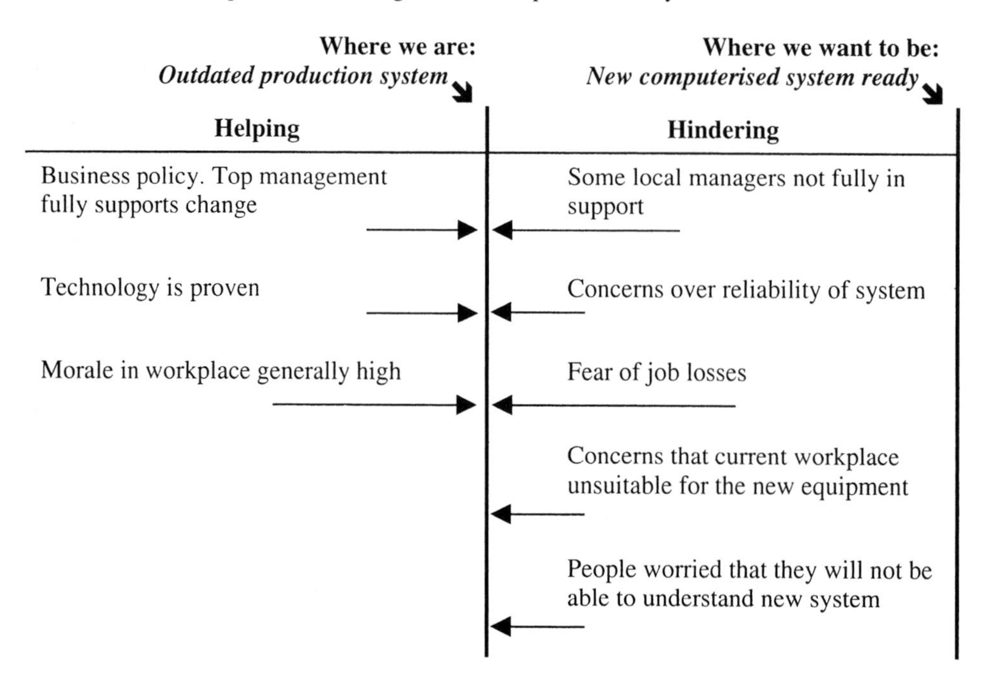

Figure 12.9 Example force field analysis

12.3 IDEA GENERATION AND DEVELOPMENT

This section describes tools that are designed to enable the creation of new ideas and/or development of these. It is common for these to follow on from some form of immersion, warm-up or problem analysis activity so that people are mentally prepared for creative work.

The striking feature of these techniques is that many of them involve the use of 'hands-on' activities. These allow people to visualise their ideas by making things, including drawings, collages or models. In Chapter 10, Liz Sanders and Colin William describe how these types of techniques help *bisociation* and *expression* – i.e. the bringing together of elements in a new way, and then expressing those new ideas for others to see and understand.

The generation of ideas is linked closely with the assessment or evaluation of them. Typically the ground rules of any idea generation method require people to abstain from criticism or rational assessment during the creative phase. Once the creative phase has been completed, the outputs lend themselves to scrutiny from others, and this can again lead to the development of further new ideas.

One of the methods in this section (see Section 12.3.2 Nominal Group Technique) incorporates concept assessment within it. Other assessment methods are, however, described in Section 12.4 – Concept Evaluation.

12.3.1 Brainstorming

Brainstorming is an idea-generating technique that is relatively quick and easy to use. Participants in the session call out ideas so that each person has the opportunity to build on the thoughts of others. The ideas are written down on a flip-chart by the moderator/facilitator as they are called out. Brainstorming is often an integral part of other group techniques, for instance, Cause and Effect Analysis (see Section 12.2.1) and Nominal Group Technique (see Section 12.3.2). Typically, brainstorming sessions last for no more than 30-40 minutes, and are often much shorter.

The moderator/facilitator describes the problem or situation for which the ideas are sought and ensures that everyone understands the objectives. The brainstorming itself can then take place. There are different ways of running brainstorming sessions. These include:

- group members call out their ideas spontaneously. This is the most familiar brainstorming method;
- the facilitator/moderator asks each group member in turn for a suggestion. If during a round a group members does not have an idea to offer, they can 'pass' until the next round (silences during sessions are allowed – they enable participants to think);
- group members spend a short time (approximately five minutes) writing down their ideas. The facilitator/moderator then asks each person in turn to read out one of their ideas, continuing round the table until all have been presented;
- group members spend five minutes writing their ideas down on paper. They then pass the sheet onto the person sitting beside them. All participants continue writing ideas, using previous suggestions in the lists as inspiration. The process is repeated until the sheets have passed around everybody, or when there are no new ideas. This is sometimes called *brainwriting* (Higgins, 1994);
- ideas are submitted on note-cards rather than called out. This enables inputs to be more easily sorted or categorised as required.

Whichever method is used, it is important that all taking part understand the basic rules of brainstorming to help create an informal atmosphere where participants feel unconstrained. There are many variations to these rules, but they can be summarised as follows:

> *Do not evaluate or criticise ideas during the session.*
> *Encourage wild or seemingly irrelevant ideas.*
> *Build on the ideas of others.*
> *Strive for quantity.*

There are many ways to encourage and enable participants to be creative during brainstorming sessions. These are explored at length in Higgins (1994) and Clegg and Birch (1999). A selection of these techniques is summarised below.

Inverse brainstorming. Instead of generating ideas for how to solve the problem, think of ways to make it worse. This can bring a fresh perspective from which real ideas for solutions can emerge.

Analogy. Identify some of the needs of the topic of the brainstorming session, then identify some examples of these needs in other areas. These areas are not restricted to products or systems – they could be industries, sports, animals and so on. Then identify the ways that these might provide direct solutions, or use them as a springboard for more abstract thinking.

'It's silly' brainstorming. The only criterion for this is that the ideas must be lateral. Anything remotely rational must be discounted. From the list of outrageous suggestions, there may be a foundation for a novel but workable solution.

Ignore physics. This is a useful technique for product development. When creating ideas, ignore the constraints of a fundamental law of physics. This may lead to the identification of solution concepts that can subsequently be made to work with appropriate development.

Word association. When inspiration is lacking, it can be helpful to generate words apparently at random so that new connections can be made, possibly leading to new solutions. A common way of doing this is to pick a word at random (using a dictionary or similar source) as a starter. The group members then call out words associated with the starter word. These can then be related back to the problem at the end to explore possible solutions. As a variation, words which 'pop out' by themselves with no apparent relationship to the starter word can be used.

Brainstorming is undoubtedly a quick and effective way of harnessing the creativity of the participants, and exploits the synergistic effects that occur when people are brought together for group discussions. To be successful, however, the group members have to participate fully. The outcomes will be disappointing if individuals feel constrained in any way. In situations where group members may not be fully at ease, it is wise to leave brainstorming until later in the session, perhaps following on from one of the immersion and warm-up activities described in Section 12.1.

Main advantages:

- relatively uncomplicated technique;
- requires little pre-preparation work;
- generates ideas quickly.

Main disadvantages:

- requires full participation from all group members;
- will not work well unless an informal atmosphere is created first;
- not necessarily exhaustive.

12.3.2 Nominal Group Technique

Nominal group technique is a structured process for generating ideas and producing immediate results in the shape of a list of rated priorities. It is suitable for groups of between six and twelve people. The procedure (from Higgins, 1994) is as follows.

Generation of ideas. Participants are given a specified time (usually five to ten minutes) to write down their ideas in response to the stated problem.

Recording of ideas. The facilitator/moderator asks participants to read out their ideas in turn, one idea at a time, and writes these on a board or flip-chart. This continues in a round-robin fashion until all the different ideas have been read out – duplications of the same idea should not be added to the list. Asking for contributions in turn in this way de-personalises them and helps emphasise equality of ideas.

Clarification of ideas. Each idea in the list is briefly discussed so that everyone understands what it is. The purpose of this step is to ensure that everyone knows what the choices are, and is not to be used as an opportunity to sell the ideas to others.

Voting on ideas. The purpose of this stage is to narrow down the list of ideas and prioritise them. This is done using a ballot process. There are several ways of managing this but, commonly, the participants are asked to write on a card their top five choices (in order of preference) and hand it to the facilitator/moderator. The results can then be tabulated to show the total number of votes each idea has received and the ranking scores. Ranking scores are determined by using a 1 to 5 scoring system; 5 points being awarded to an idea whenever it is ranked top. If necessary, the voting process can be repeated to further refine the output, but on the second iteration people should be asked to write down just their top three ideas. On the second iteration, ranking scores are determined using a 1 to 3 scoring system.

The use of a structured approach, especially the incorporation of a voting system, is attractive because it ends with some form of result, which is often satisfying for participants. For some participatory design projects, particularly those where the group has decision-making authority, formal outputs such as this may be a key requirement so that an audit trail is created.

Main advantages:

- requires little pre-preparation work;
- process encourages all to take part;
- minimises impact of dominant group members;
- produces a tangible output, i.e. a prioritised list of ideas or actions.

Main disadvantages:

- some participants may feel pressured to contribute;
- structure may be inhibiting;
- voting process takes up time during the session.

12.3.3 How–how Diagrams

This method is particularly useful for seeking ways of refining and implementing solutions to problems. It builds on other techniques described here (for example, Section 12.3.2 Nominal Group Technique) by taking the agreed solutions and then exploring the details for how to make them work.

How-how diagrams use the same principles as why-why diagram*s* (see Section 12.2.2), but in reverse. Starting with a high level statement of a possible solution, the methods for achieving the solution are progressively elaborated by continually asking the question 'how?' (see Figure 12.10).

Main advantages:

- uncomplicated but powerful technique to identify how an idea or solution can be developed and implemented;
- forces participants to think about practical issues of implementation;
- harnesses the collective knowledge and skills of the participants.

Main disadvantages:

- focus on implementation and practicality may reduce creativity;
- less suited to focus groups with consumers, who may be unfamiliar with the method.

Objective	HOW?	HOW?
Make it easier to replace the cartridge	Make it easier for user to find the cartridge	Give clear instructions on the display
		Provide a mimic diagram on the display
		Clearly label the cartridge
		Colour code the cartridge
	Make it easier to remove the old cartridge	Make the handle obvious – shape or colour
		Provide bigger finger grips
		Reduce force required
	Ensure user removes locking tag	Redesign tag so cartridge cannot be fitted with it on
		Provide a big warning label
	Ensure the user does not fit the cartridge upside down	Redesign the aperture so that it only fits one way
		Provide a warning label
	Make sure new cartridge locates correctly	Change design of runners so that cartridge is guided into place
		Provide a noticeable 'click' so they know it is in
	Provide a better user guide	Reduce detail in the graphics
		Use colour
		Number each step

Figure 12.10 Example how-how diagram

12.3.4 Drawing

Drawing can be an extremely versatile and powerful way of enabling individuals to generate and communicate ideas. When people draw their ideas, the outputs often contain details and concepts that are not easily described verbally. Sometimes they contain information that the participants have taken for granted and have not thought worthy of mention as part of the discussion. The completed drawings are a valuable stimulus for further discussion and elaboration or development of ideas. They are also extremely helpful when presenting findings to clients, or to support designers with subsequent product or system development. Drawing activities are fun, stimulating and thought provoking.

Drawing exercises can be carried out with people working individually or together in small groups. They can be used in many different ways and for different purposes. For example, in a session on kettle design, participants could be asked to draw their 'ultimate' (future) kettle. In a session looking at improving a workplace, people could be asked to create two drawings, one of their existing workplace (helping to define the main problem areas) and the second to show how it might be re-designed. For a website, people could be asked to draw out a chart showing the key chunks of information and the links and hierarchies between them. Any type of drawing or representation can be used, including plans, 3D sketches, computer screen layout designs or charts. It is often helpful if the participants add their own personal comments to the drawings to highlight key features (see Figure 12.11).

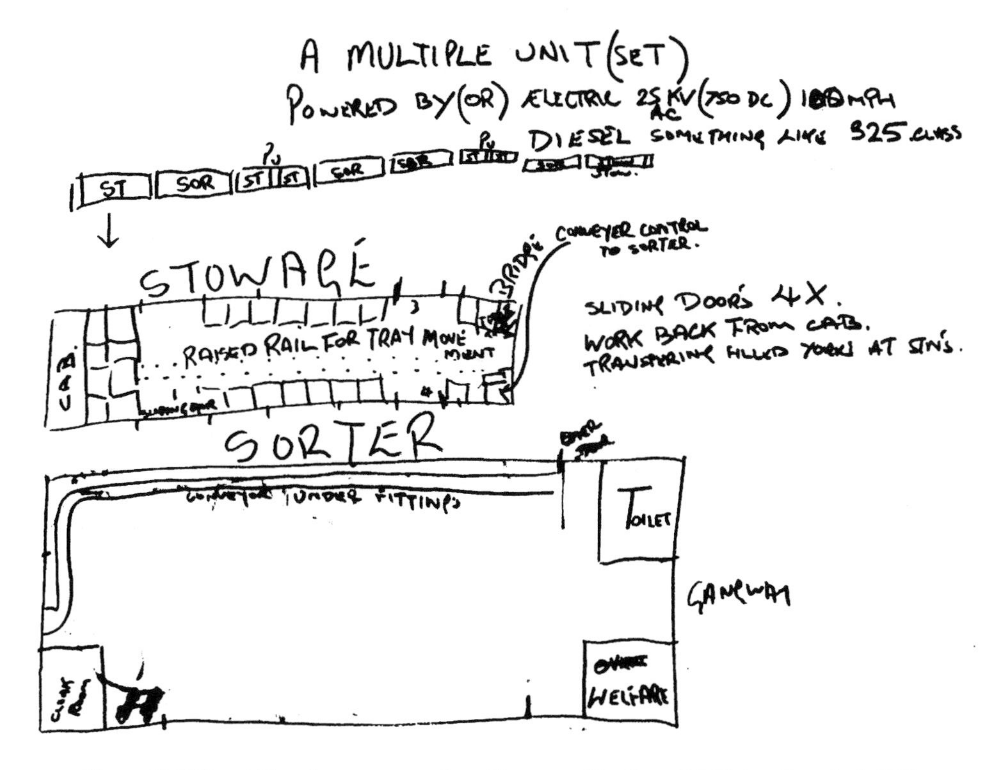

Figure 12.11 Users' ideas for a new type of travelling post office (from Rainbird and Langford, 1998)

The only materials required for drawing exercises are drawing media and surfaces (paper and pens). Large sheets of paper (up to flip-chart size) take up space on the table, but provide space to enable drawings to develop. They are also easier for people to see from a distance during feedback or subsequent discussion. It is helpful to provide several pens and pencils in a variety of colours. This allows people to make their drawings clearer and more informative by highlighting different features and using colour coding where required.

To get the best from drawing exercises, participants need to be focused on the issues under discussion and also to be fully settled in. It is usually best to schedule such exercises later in the session after some of the other activities have taken place, for example, one of the immersion and warm-up techniques described in Section 12.1. It is also helpful to provide a briefing for the activity to ensure participants know what the objectives are, and also the constraints (if any). In some cases, it can also be helpful to remind people of the key issues to consider, for example, types of user, types of use and features required. For more complex tasks, it is helpful to write the briefing down. Often, the participants will be unfamiliar with this type of activity, so it is important to make them feel at ease, for example, by reassuring them and advising them not to worry about the quality of their drawing skills, or whether their annotations are spelt correctly.

It can be useful in some cases to issue slightly different briefings to individuals or groups. For example, some individuals could be asked to draw solutions taking account of constraints such as cost or technical feasibility, whereas others could be asked to take a 'blue sky' approach.

The time required for drawing activities is dependent on the complexity of the task set and the desired outcomes. In some cases, time may be limited to five or ten minutes to ensure that participants concentrate on just the key features. In other cases, it is better to provide more time (up to about 45 minutes) to allow ideas to be explored and developed.

Main advantages:

- relatively uncomplicated and straightforward – requires minimal pre-preparation;
- transcends limitations of words by using drawings to help articulate thoughts and ideas;
- provides a stimulus for subsequent discussion and probing – acts as a vehicle for generating more specific information and design ideas;
- outputs can be used to help present findings and to provide design rich data for designers.

Main disadvantages:

- requires reasonable amount of time in the session to complete the activity (typically between 5 and 45 minutes);
- unfamiliar exercise – may be uncomfortable for some participants. Clear briefing and some assistance from the moderator/facilitator may also be required.

12.3.5 On-line Illustrator

The use of an on-line illustrator is a variation to drawing (see 12.3.4). A professional artist, illustrator or designer can rapidly translate spoken thoughts into sketches, thus visualising the ideas on behalf of the creators. Once created, the drawings can then form the focus for the development of the basic ideas, leading to improvement, or to completely new ideas. O'Brien (1981) used this method as part of his Shared Experience Events. A similar technique was used by industrial designers Seymour Powell during the television series *Better by Design* (Channel 4, 2000).

The originators of the ideas may perceive that they lose control to some extent because the realisation of their thoughts is in the hands of someone else. However, the whole group can contribute more easily to the development of the idea as the picture gradually builds up. Also, participants are not constrained by their limitations as artists, or lack of confidence due to a perceived lack of drawing skills.

Main advantages:

- transcends limitations of words by using drawings to help articulate thoughts and ideas;
- some participants may be more comfortable with a third party doing the drawing;

- provides a stimulus for the development of ideas by the whole group;
- outputs can be used to help present findings and to provide inspiration for designers.

Main disadvantages:

- requires reasonable amount of time in the session to complete the activity (typically between 5 and 45 minutes);
- some individuals may not be able to contribute to the development of the ideas;
- additional cost of hiring an illustrator.

12.3.6 Storytelling

Storytelling is a method for presenting ideas. Participants are given a problem to solve, as they might be with a drawing exercise, but are asked to present their ideas in the context of a short story. This provides context and brings the ideas to life. There are some similarities to scenario-based discussion (see Section 12.4.1) and simulation/role-playing (see Section 12.4.5).

Kuhn (2000) describes a study where focus groups were a key part of the development of requirements for a home automation system. As part of this, participants were asked to develop ideas for how particular features of the system could be useful and how they could work. These features included time scheduling, shopping lists, reminders and regulation of the heating system. Small groups of participants were given a clear briefing of what they had to achieve and were asked to work out and present two or three ideas in the form of a short story.

Main advantages:

- relatively uncomplicated – requires minimal pre-preparation;
- provides a context for the explanation and increases understanding of ideas;
- provides a stimulus for subsequent discussion and probing;
- outputs can be used to help present findings and to provide inspiration for designers.

Main disadvantages:

- requires considerable amount of time in the session to complete the activity;
- unfamiliar exercise – may be uncomfortable for some participants. Clear briefing and some assistance from the moderator/facilitator may also be required.

12.3.7 Two-dimensional (Layout) Modelling

Two-dimensional modelling involves the positioning of objects on a flat surface to explore and test ideas for design problems primarily concerned with *layout*. Examples where this type of approach is applicable include:

- floor layouts for buildings or workplaces;
- layouts of equipment on workstations;
- placement of the key elements in a vehicle cockpit;
- positioning of controls and displays on an instrument panel;
- positioning of information on display screens or panels.

Layout models allow many issues to be tested. With buildings and workplaces, for example, these include circulation space, product flows, access and people or vehicle movements. For workstations, issues such as vision, reach and space can be assessed. For display and control applications, issues such as grouping of information elements and sequence of use can be explored.

Because it is relatively quick and easy to do, layout modelling may be used as a stepping stone prior to the further development of ideas in three dimensions. For example, a range of ideas for the positioning of features in a vehicle cockpit might be generated using layout modelling, enabling the best two or three to be identified for detailed development in three-dimensional mock-ups.

In some cases, layout modelling is best carried out in scale, for example, the positioning of rooms or areas in a building. In other cases, however, working in full scale is more helpful, for example, the placing of equipment on a workstation such as a supermarket checkout. Full-scale modelling of this nature is a relatively quick and simple way of bringing things to life for the participants.

Scale layout modelling can be done simply using paper or card cut-outs of the components (for example, equipment or workstations) and laying these out on paper or plans. As described in Snow *et al.* (1986), overhead projections of designs can also be used. For full-scale modelling, cut-outs can be laid out on the floor or on a large table – although care should be taken that the shape of the table does not constrain the designs. To aid visualisation and testing, it is sometimes more helpful to have simple three-dimensional card mock-ups of key items (Wilson, 1991). For example, with a supermarket checkout, this would include data entry keyboard, display panels, weighing scales, printer and cash drawer.

Within discussion sessions, layout modelling exercises are best carried out with people working in groups of two or three. Between 15 and 45 minutes are required, depending on the complexity of the task set. As with drawing exercises, participants should be warmed up and focused on the key issues prior to starting. Again, it is important to brief people properly. When the exercise is complete, the groups present their findings to the remainder of the participants, providing further stimulus for discussion and probing of issues.

Main advantages:

- useful for testing functional issues relating to workplaces, workstations and systems;
- intuitive activity – participants find it enjoyable, creative and interesting:
- layout models help visualise ideas and provide a stimulus for discussion and probing;
- quick method for idea generation and testing. Can be used as a stepping stone to eliminate poor ideas prior to further development in three dimensions;
- outputs can be used to help present findings and to provide inspiration for designers.

Main disadvantages:

- requires considerable amount of time in the session to complete the activity (typically between 15 and 45 minutes);
- full-scale layout modelling requires more space and materials;
- not useful for exploring emotional aspects of design such as appearance. Unlikely to be useful for consumer products.

12.3.8 Three-dimensional (Form) Modelling and Mock-ups

For some topic areas, three-dimensional modelling provides an excellent opportunity for people to express and to test their ideas. By building models and mock-ups people can:

- generate and assess ideas for the three-dimensional form of a product;
- assess issues relating to size, shape and comfort;
- test ideas for positioning controls and/or displays on an object;
- generate and assess ideas for positioning of items in three-dimensional spaces, for example, workstations or aircraft cockpits;
- generate and test entire space layouts, for example, workstation layouts in a control room.

Three-dimensional modelling can be carried out at reduced or full scale. The choice will depend on the application and/or the facilities available. Smaller, handheld objects lend themselves readily to full-scale models, whereas larger sized projects such as vehicle cockpits require more space and materials. Where it is possible, however, using full-scale modelling of larger sized objects is beneficial as it allows participants to become part of the model and to carry out simple simulations or role playing exercises. The crane driver example in Chapter 11 is an example of this.

There is an almost limitless number of ways of constructing three-dimensional models. However, practicality is the key when designing modelling exercises to be carried out as part of group discussions. Issues to consider include the following.

Ease of use. It is important the method is intuitive for the participants and requires no special skills to use. To a certain extent, this is dependent on the types of people taking part. For example, use of nuts, bolts and simple tools may be acceptable for engineers or product designers, but not for some other user groups.

Speed of use. It must be possible for people to complete the exercise in a reasonable time – typically between 15 and 60 minutes depending on the application. There is no requirement for the outcomes to be sophisticated or finely detailed. The aim is simply to enable quick generation, development and/or testing of ideas.

Protection of participants. It is important to ensure that participants will not be harmed or their clothing damaged by the modelling materials or tools used. Again, this will be dependent to some extent on the types of people taking part. It may be necessary to provide some protective clothing and/or facilities for people to clean themselves up afterwards.

A selection of modelling methods that have been used within group discussion sessions is given below.

Clay modelling. This is useful for smaller scale objects and can be used to explore size, shape and form. Positioning of handles, controls and displays can also be incorporated. The medium is easily worked and does not constrain the users to any particular shapes or forms. Dolan *et al.* (1995) used clay modelling during their focus groups exploring designs of telephone handsets.

Velcro-modelling kits. Liz Sanders and Colin William describe these in Chapter 11. Such kits consist of a range of shapes, buttons and other items that can be quickly and easily stuck together to create objects. The components are deliberately ambiguous in form and purpose so people are not constrained in the ways they use them.

Cardboard modelling. Cardboard is an inexpensive and versatile material that can be easily cut to shape and taped together to create mock-ups. If required, support structures such as Dexion can be used. More detailed features can be added by simply drawing or sticking them on. Cardboard modelling is particularly suited to larger scale mock-ups such as workstations (e.g. Wilson, 1991) or vehicle cockpit mock-ups. Assistance may be required with some user groups, for example when cutting the card to shape.

Use of existing components. With full-scale mock-ups it is sometimes possible to incorporate existing components such as seats, keyboards or control panels. Sometimes it is more effective to provide mock-ups or models of these components to reduce weight or minimise risk of damage. The integration of real components helps to add realism to the model and can help with the testing of concepts.

Main advantages:

- valuable for idea generation and testing, including visual appearance and three-dimensional form of objects;
- intuitive activity – participants find it enjoyable, creative and interesting;
- models help visualise ideas. With full-scale models, participants can carry out simple simulation exercises and gain a more realistic appreciation of key issues;
- provides a stimulus for subsequent discussion and probing;
- outputs can be used to help present findings and to provide inspiration for designers.

Main disadvantages:

- requires considerable amount of time in the session to complete the activity (typically between 15 and 60 minutes);
- requires appropriate materials and possibly tools and/or protective clothing. Participants may require skill and/or assistance using tools and materials;
- full-scale modelling requires more space.

12.3.9 Design Decision Groups

The primary use of Design Decision Groups is to facilitate the participation of individuals (e.g. users, consumers, customers and stakeholders) in the design of changes to the equipment and workplaces they use. They do this by incorporating a range of techniques such as word maps, drawing, discussion and modelling within a pre-determined structure or process. This leads the participants through various phases including problem analysis, idea generation and concept evaluation. Depending on the specific topic of interest, the process usually spans two or three sessions (see Figure 12.12). The groups typically involve six to eight participants, but can work with more or less people if required.

Design Decision Groups have been used for many purposes including the design of retail checkouts, telephone exchanges and library issues desks (Wilson, 1991). Other examples are given in Chapter 11.

Main advantages:

- provides a 'complete' approach for participatory design, incorporating problem analysis, idea generation and concept evaluation;
- adaptable and flexible – can be used for many different purposes;
- provides stimulus for probing to enable detailed understanding of key issues;
- outputs can be used to help present findings and to provide inspiration for designers.

Main disadvantages:

- requires considerable amount of time – each session lasts from two to four hours;
- for participatory design applications, it is sometimes difficult to get participants released from work for sufficient time periods;
- modelling exercises require space and materials.

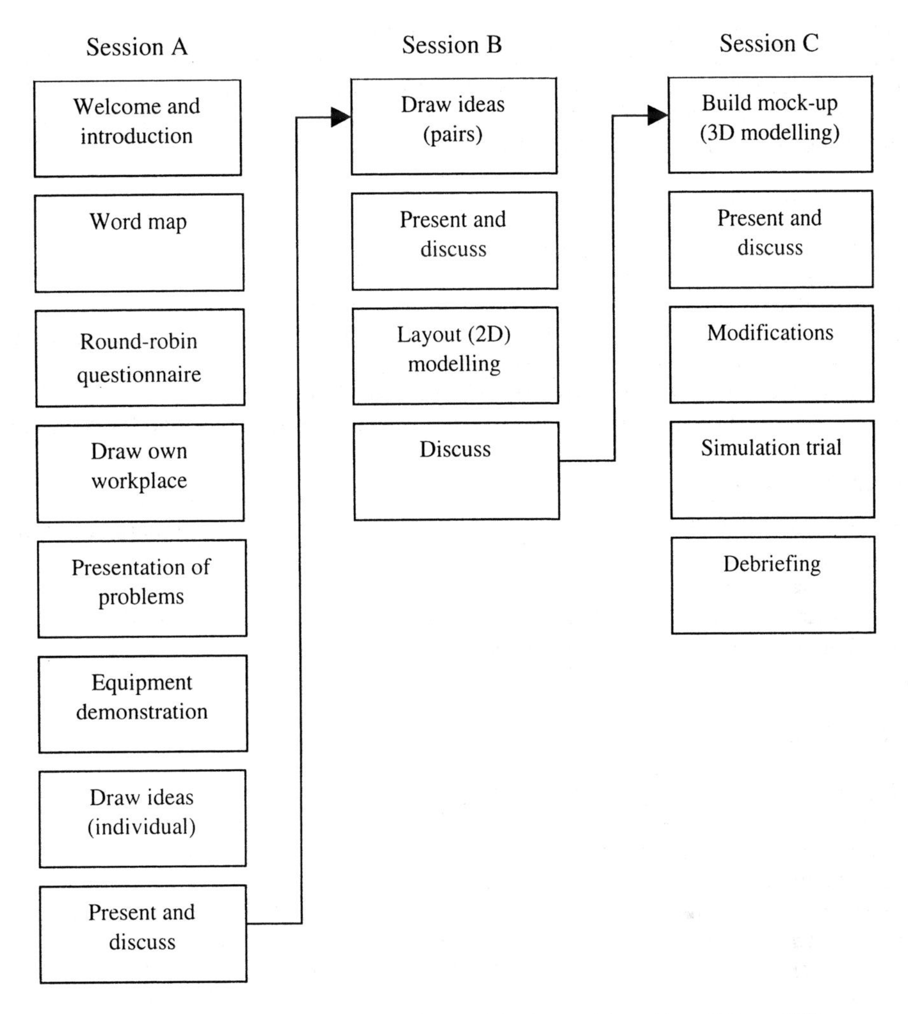

Figure 12.12 Example of a process for a Design Decision Group (adapted from Wilson, 1991)

12.4 CONCEPT EVALUATION

The techniques described in this section are designed primarily for the evaluation of concepts, products or ideas. In some cases, a focus group session will be set up purely with this aim in mind, for example, to gain consumer feedback on new designs for products or systems. In other cases, however, the techniques may be required for the assessment of ideas resulting from the session, following on from one of the idea generation techniques described in the previous section.

Some of the methods described here are designed to present ideas and concepts to the participants, for example, visual evaluation (see Section 12.4.2) and scenario-based discussion (see Section 12.4.1). These promote meaningful discussions, with feedback emerging in an informal way. Other methods tend to be more formal, for example, questionnaires (see Section 12.4.10). These allow data to be collected in a systematic way, permitting further statistical analysis if required. Other methods are more analytical, for example, decision-making analysis (see Section 12.4.11). These provide systematic approaches, taking the participants through logical steps to support decision-making.

Some of the methods described in earlier sections also incorporate evaluation methods. These include nominal group technique (see Section 12.3.2) and force field analysis (see Section 12.2.3).

12.4.1 Scenario-based Discussion

The principle behind scenario-based discussions is that by describing usage scenarios, people are forced to think about what they do, when they do it and how. This helps raise their awareness, including any problems or good features of existing products and systems. In some ways, this is similar to the immersion and warm-up exercises described earlier, for example, diaries/journals or 'day-in-the-life' exercises (see Section 12.1).

Scenario-based discussions are particularly helpful when the topic area is concerned with the future, rather than just the 'here and now'. People can relatively easily explain and discuss their feelings and opinions about the present. It is more difficult, however, for them to imagine how they might feel about future products or systems, or to define what they might require from them in the future. The use of scenario-based discussions provides a structure that helps address these problems.

Lee Cooper and Chris Baber describe scenario-based discussions in detail in Chapter 9. A basic procedure for sessions to explore future products and services is as follows.

Develop usage scenarios. Ask each participant to develop a set of usage scenarios for the current day equivalent of the future product or system. For example, Lee Cooper and Chris Baber used ATMs (automated teller machines, or cash-points), Internet banking and wallets or purses as current day equivalents of the electronic wallets of the future. Give the participants pens and paper to record their scenarios, using pictures and words.

Identify significant features. Ask each participant to identify features of the scenarios that they consider significant. For example, with the wallet project described above, storage and access were seen as significant features.

Discuss features. Ask the group to discuss the pros and cons of the significant features. Use the discussion session to elaborate these, allowing participants to use their own stories and analogies to bring them to life.

Demonstrate new product concept. Introduce the new product or service to the group, explaining its main features.

Assess concept. Ask the group to assess how the features of the new product or system would emphasise or de-emphasise the pros and cons identified earlier.

Although the procedure described above is for assessing a previously identified product or system (for example, an electronic wallet prototype), it can also be used to identify requirements for new products and systems. In this case, the fourth step (demonstrating the new product) is replaced with an *idea generation* step where participants are asked to identify new features that will emphasise the pros and de-emphasise the cons identified earlier.

Main advantages:

- helps when evaluating future products or systems;
- can also be used as a basis for generating and evaluating new ideas;

- provides a stimulus for discussion and probing;
- outputs can be used to help present findings and to provide inspiration and stimuli for designers.

Main disadvantages:

- fairly complex procedure;
- may be unfamiliar for some participants. May require clear briefing and some assistance from the moderator;
- may be time-consuming.

12.4.2 Visual Evaluation of Products or Systems

The simplest form of product or system user testing is to provide a visual stimulus, which subsequently allows the participants to provide feedback in the discussion and also, if required, via a more formal mechanism such as a questionnaire.

The method is versatile and can be employed to assess existing or future products and systems. It is particularly helpful for future products and systems because key details on a wide range of options can be presented or demonstrated to the participants without the need to produce expensive prototypes or models.

Product or system demonstration for this form of evaluation can take many forms, including:

- visuals or images showing how a product looks, what its key features are and how it is intended to be used (Figure 12.13);
- schematics or storyboards showing a proposed system, incorporating its key features and how it operates;
- computer-generated animations to show a product from a range of different angles, or with different surface colours or features;
- virtual reality models to show the layout of rooms and workstations in a building and allow a virtual tour;
- screen shots of a proposed human computer interface to show appearance, general operating principles and system features. Examples of typical interactions can also be shown.

Whilst these are relatively inexpensive and simple ways of presenting concepts to participants, the feedback will be of limited quality. Without 'hands-on' experience it is hard for individuals to gain a realistic impression of how something really looks and feels, or whether it will in reality provide the benefits they are looking for.

Visual presentation of ideas does, however, provide a valuable stimulus for discussion and enables a wide range of different concepts to be evaluated early in development. It is also provides a reasonably realistic simulation of Internet or catalogue shopping, where products cannot be handled or 'tried out' and have to be assessed largely from their looks alone.

In Chapter 3, Wendy Ives describes how *concept boards* can be used to introduce new concepts and ideas to the participants for feedback. She recommends that these are designed to be easy to read and understand, with illustrations that are straightforward to interpret. This is vital for sessions with older participants and those who may be visually impaired – see Chapter 7 and Appendix 1 in Chapter 5. Interestingly though, Wendy Ives also notes that it can be helpful if concept boards have a home-made or unfinished look so that people are encouraged to suggest alterations.

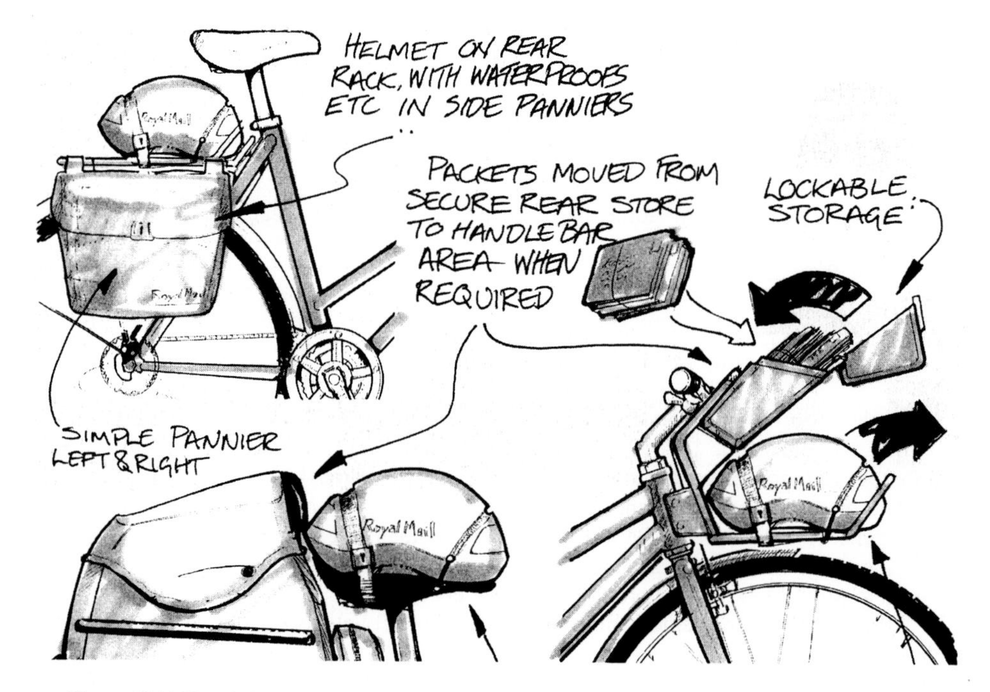

Figure 12.13 Visual stimulus to gain feedback on ideas for mail delivery cycles (courtesy of Royal Mail)

Main advantages:

- adaptable and flexible. Can be used for a wide range of products and systems, and for evaluating future products or systems at an early stage in the development process;
- simulates Internet or catalogue purchasing experience;
- relatively low cost;
- provides a stimulus for discussion and probing.

Main disadvantages:

- quality of feedback is limited due to lack of hands-on experience;
- presentation/concept boards are time consuming to prepare.

12.4.3 Product Handling

Product handling enables participants to handle objects in much the same way that they do in a retail showroom. Whilst this does not permit any realistic testing (for example, filling a kettle with water), it does allow people to form *gut reactions*, taking into account weight, comfort and texture as well as physical appearance. Participants can also assess ease of use to some degree, and gain an impression of features they like or dislike, for example, the *feel* of the way the switch operates. The method can be used with real products (Figure 12.14), or models and prototypes of future products. As with visual evaluation (see Section 2.4.2), feedback can be via the discussion or more formally, using questionnaires for example. Observation (perhaps with video recording) of participants whilst handling the products can also provide additional information to the researchers about participants' reactions and some of the more obvious usability problems.

Figure 12.14 Product handling exercises

Product handling provides a much better basis for participants to provide feedback on products, and can be more effective than relying on evaluation based on visual images alone. It simulates the situation people usually face when purchasing in a retail environment. It has limited use for systems or for complex products, which are likely to require more in-depth assessment, perhaps using some form of user testing (see Section 12.4.4). Some simple guidelines for setting up a product handling exercise are given below.

Labelling. Ensure that all the products are adequately labelled, using numbers or letters.

Presentation. Make sure they are presented in a convenient way, for example, on a table at an appropriate height.

Briefing. Provide a clear briefing to participants so that they know what to do and in what order, including completion of questionnaires or forms if appropriate.

Minimise bias. Ask participants not to discuss their thoughts before they have completed the exercise. If appropriate, remove or disguise branding and/or manufacturer information – this may influence some participants' views.

Main advantages:

- useful for evaluating existing products or models/prototypes of products under development;
- feedback is based on hands-on experience, so is more realistic than purely visual stimulus;
- simulates a typical consumer purchase in a retail environment;
- provides a helpful stimulus for discussion and probing;
- observation and questionnaire feedback can provide additional information.

Main disadvantages:

- quality of feedback is limited – handling of products is not truly representative of day-to-day use and experience;
- more suited to the evaluation of products than systems;
- requires substantial planning.

12.4.4 Product and System User Testing

There is a fine dividing line between product handling (see Section 12.4.3) and user testing. It could be argued that product handling is a form of user testing in that people get the chance to handle the product and imagine what using it might be like. The distinction drawn here is that with user testing, participants engage in a meaningful interaction with the product or system that attempts to mimic real-life usage. For example, rather than just handling an iron, participants might be given the chance to fill it with water, iron some clothes and then pack away afterwards.

In some cases, users cannot give meaningful feedback without having taken part in some form of user testing. For example, people could not discuss the ease of use of a new photocopier if they had not had the chance to carry out a range of copying jobs, filling it with paper or clearing a jam. Likewise, a focus group discussing the design of a new website will be limited in value if the participants are only given a demonstration. The feedback will be much more useful if they are given a chance to try the system out, for example, by performing a range of tasks on a working prototype.

Apart from providing the participants with the necessary experiences for the discussion session, user trials can also provide objective data. This includes information such as the time to complete certain tasks and the number of errors made. Observation of individuals carrying out the tasks can also provide helpful supplementary information. Questionnaires or interviews may be used to capture subjective feedback. The validity of this data is, of course, dependent on a number of factors such as the number of participants and the realism of the trial, but is nevertheless often useful to support the focus group findings or vice versa. Caplan (1990) describes how he *twins* focus groups with user trials for this reason.

User testing can be associated with focus groups in two main ways, either by incorporating it as an exercise within the session, or by carrying it out in advance. The choice will largely be dependent upon practicality and appropriateness. With lengthy or complex trials it is best to carry them out in advance, and use the focus group to discuss the participants' experiences and the results. Quick and simple trials requiring limited facilities can, however, be incorporated within the session, particularly if all the participants can carry out the trial at the same time.

This chapter is not intended to describe in detail the different methodologies and techniques for user testing. Some general principles for the design of user trials when used alongside focus groups are however described below. For further guidance in this area, refer to McClelland (1995) and Nielsen (1997).

Define the purpose of the trial. This is likely to be similar to the overall aims of the focus group study. This will determine issues such as which features or attributes of the product or system need to be assessed and whether it is necessary to compare two or more similar products or systems.

Acquire product(s) for the trial. This will be easy if off-the-shelf items can be used, but in many cases, production of a mock-up or prototype with the necessary attributes and/or functionality will be required. Often such mock-ups can be fairly crude so long as they simulate the desired features or attributes effectively enough. For example, with a human computer interface, screens can sometimes be presented on paper, or as simple slide shows.

Define appropriate trial tasks. Tasks determine the interactions that the participants have with the product or system. If the aim of the study is to uncover and increase the understanding of participants' general reactions, then the tasks need to expose people to a wide range of interactions. If the study is concerned with more specific details, then the tasks should be focused more on those areas. The tasks should simulate real-life experiences as far as is practicable. In cases where off-the-shelf products are being explored, it is sometimes possible to give them to individuals to use as part of their normal daily routines prior to the group session.

Define appropriate measures and data collection methods. These include observation (possibly with video recordings), time taken to perform tasks and errors made. Subjective data such as comfort, satisfaction and ease of use can also be collected using questionnaires or interviews as appropriate. Bear in mind that the subjective issues raised may provide a key focus for the subsequent discussion session.

Define a procedure for the trial. This should include an introduction, familiarisation, trial tasks and data collection. Test the procedure in advance to establish the time required and to ensure smooth running.

Minimise potential for bias as far as possible. Ensure that participants do not share their views and experiences before the trial has been completed. If several different products or systems are being tested one after another, ensure that the sequence of testing for each person is different. This helps minimise the effects of learning (it tends to get easier as they go on) and tiredness (it becomes more difficult). Be consistent with the treatment of each participant. Employ the same script and procedure for each participant to reduce the possibility of people picking up unintentional cues that might affect their performance or responses.

Main advantages:

- helpful for evaluating existing products or models/prototypes of products under development. Can also be used for systems as well as products;
- feedback is based on experience performing *real* tasks, which is more realistic than purely visual evaluation or product handling;
- provides a stimulus for discussion and probing;
- observation and questionnaire feedback can provide additional information.

Main disadvantages:

- quality of feedback is limited to some extent – even well-designed trials are not truly representative of day-to-day use and experience;
- trials can be time-consuming to prepare and conduct.

12.4.5 Simulation and Role Playing

Within a focus group or participatory design context, simulation and role playing is a way of bringing concepts to life so that they can more easily be assessed. In broad terms, with simulation and role playing, the concept under scrutiny is mimicked in some way, with participants playing roles to enact typical scenarios of use. The method is extremely flexible and can be adapted to suit many different applications from exploring novel ways of interacting with a product or system, through to the layout of a manufacturing workstation or the design of new work processes.

The complexity of the simulation can be quite basic, for example, *walking through* a series of tasks on a mock-up workstation built by the participants as part of a modelling exercise (see Section 12.3.8). In some cases, however, a more sophisticated approach might be needed, using technical support, or well co-ordinated back up teams.

Sato and Salvador (1999), for example, use *focus troupes* in which short skits featuring new product concepts are acted out for the participants, demonstrating how the new products might be used. This provides a full understanding of the operations and implications of the product. The participants are then able to take part in structured conversations about the concept to test understanding, explore possibilities and provide criticism. Sato and Salvador also describe a range of other theatre techniques to enable the participants to become more involved. These include the following.

Act out roles or skits. These can be scripted or unscripted, and can be acted out individually or in groups. This can support researchers' understanding into how individuals structure their ideas and experiences.

Act out what goes on inside a product. This helps reveal participants' beliefs about a product. It can help generate ideas for icons and make processes more visible.

***Transform* a product.** One person acts out using the product, then hands it on to the next person who builds on this, or finds a new use. This reveals stereotypical actions and can sometimes lead to novel uses.

With *behaviour anticipation* (McCallion, 1999), scenario planning and rapid prototyping are used to define a range of possible usage scenarios for a new product or system, and then immerse potential users in those scenarios. The process starts by analysing current processes, practices and environments, including interviewing potential users or users of existing equivalent products. This information is then used to develop a range of discrete scenarios to explore key features of the new product or system. The scenarios are described in detail through written stories, incorporating models of different options for the product or system. Potential users then act out the scenarios, interacting with the prototypes as required. This exposes them to a wide range of possibilities, and helps them provide more useful feedback, including identification of their own preferred scenarios.

Main advantages:

- useful for evaluating new concepts for products or systems, including future concepts where there is no experience of use;
- involving participants in the exercise, provides a form of hands-on experience;
- provides a stimulus for discussion and probing;
- video recordings of exercises can provide valuable feedback for clients or designers.

Main disadvantages:

- quality of feedback is limited to some extent – even detailed simulations are not truly representative of day-to-day use and experience;
- exercises take time to set up and run;
- sophisticated simulations require extensive detailed planning and support.

12.4.6 Product Personality Profiling

Product Personality Profiling is a projective technique that provides an insight into participants' emotional responses to a product, including *who* they perceive to be the target user. It is particularly geared towards products, but could potentially be adapted for use with other applications; website designs for example. Feedback from such exercises helps to reveal how various features of a product might determine consumers' emotional responses to it. This can help determine design directions for new products.

With this technique, participants are asked to imagine a product as a person with a particular personality, and provide information regarding its character and lifestyle, for example, gender, age and occupation. The exercise is carried out within a short space of time (typically 2-3 minutes per product) in order to encourage rapid 'gut' responses. The results provide a stimulus for discussion to further understand the motivations behind people's choices, and can also be analysed later in more depth if required.

A practical method for carrying out a product personality profiling exercise is described by McDonagh *et al.* (2002). Participants use a questionnaire to record their responses to the different products (Figure 12.15). This can be an unfamiliar activity, which some people may experience difficulty completing. As shown in the example, providing a list of cues helps participants fill in the forms.

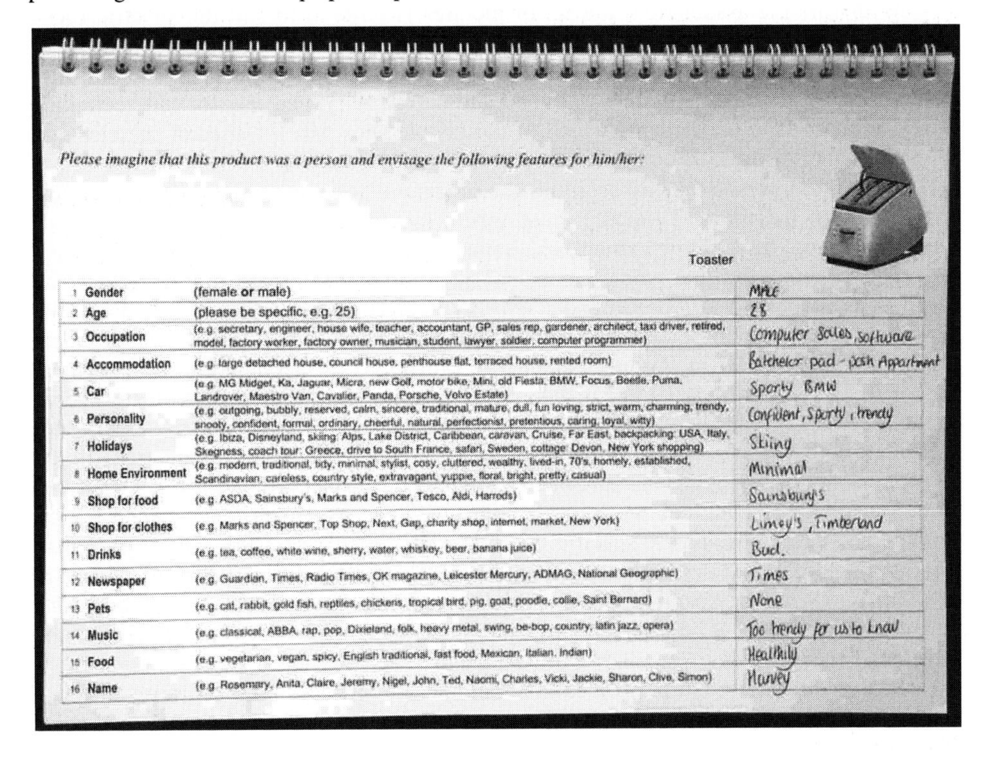

Please imagine that this product was a person and envisage the following features for him/her:

Toaster

1	Gender	(female or male)	Male
2	Age	(please be specific, e.g. 25)	28
3	Occupation	(e.g. secretary, engineer, house wife, teacher, accountant, GP, sales rep, gardener, architect, taxi driver, retired, model, factory worker, factory owner, musician, student, lawyer, soldier, computer programmer)	Computer Sales, Software
4	Accommodation	(e.g. large detached house, council house, penthouse flat, terraced house, rented room)	Batchelor pad - posh Appartment
5	Car	(e.g. MG Midget, Ka, Jaguar, Micra, new Golf, motor bike, Mini, old Fiesta, BMW, Focus, Beetle, Puma, Landrover, Maestro Van, Cavalier, Panda, Porsche, Volvo Estate)	Sporty BMW
6	Personality	(e.g. outgoing, bubbly, reserved, calm, sincere, traditional, mature, dull, fun loving, strict, warm, charming, trendy, snooty, confident, formal, ordinary, cheerful, natural, perfectionist, pretentious, caring, loyal, witty)	Confident, Sporty, trendy
7	Holidays	(e.g. Ibiza, Disneyland, skiing: Alps, Lake District, Caribbean, caravan, Cruise, Far East, backpacking: USA, Italy, Skegness, coach tour: Greece, drive to South France, safari, Sweden, cottage: Devon, New York shopping)	Skiing
8	Home Environment	(e.g. modern, traditional, tidy, minimal, stylist, cosy, cluttered, wealthy, lived-in, 70's, homely, established, Scandinavian, careless, country style, extravagant, yuppie, floral, bright, pretty, casual)	Minimal
9	Shop for food	(e.g. ASDA, Sainsbury's, Marks and Spencer, Tesco, Aldi, Harrods)	Sainsbury's
10	Shop for clothes	(e.g. Marks and Spencer, Top Shop, Next, Gap, charity shop, internet, market, New York)	Limey's, Timberland
11	Drinks	(e.g. tea, coffee, white wine, sherry, water, whiskey, beer, banana juice)	Bud.
12	Newspaper	(e.g. Guardian, Times, Radio Times, OK magazine, Leicester Mercury, ADMAG, National Geographic)	Times
13	Pets	(e.g. cat, rabbit, gold fish, reptiles, chickens, tropical bird, pig, goat, poodle, collie, Saint Bernard)	None
14	Music	(e.g. classical, ABBA, rap, pop, Dixieland, folk, heavy metal, swing, be-bop, country, latin jazz, opera)	Too trendy for us to know
15	Food	(e.g. vegetarian, vegan, spicy, English traditional, fast food, Mexican, Italian, Indian)	Healthily
16	Name	(e.g. Rosemary, Anita, Claire, Jeremy, Nigel, John, Ted, Naomi, Charles, Vicki, Jackie, Sharon, Clive, Simon)	Harvey

Figure 12.15 Product Personality Profiling (from McDonagh *et al.*, 2002)

Main advantages:

- provides insight into people's emotional responses to a product and the reasons why;
- can help determine design direction for products in development;
- provides a valuable stimulus for discussion and probing.

Main disadvantages:

- participants may be unfamiliar with the exercise, experiencing difficulty in responding and completing the task;
- very subjective and responses can vary greatly between individuals;
- interpretation of the results can be complex;
- requires considerable pre-preparation, for example, product feedback forms.

12.4.7 Association

Association is another type of projective technique. It generates information from participants by encouraging them to make associations with other stimuli as a way of expressing their feelings towards a concept, product or service or other entity. These reactions can then be used to probe further and stimulate discussion. Greenbaum (1998) describes several types of association techniques. These include *personality associations, situational associations* and *forced relationships*.

Personality associations use photographs of people to stimulate the participants' thinking and help them articulate their feelings about the topic. Personality associations can be *fixed* or *variable*.

With variable association, the set of pictures is tailored to suit the specific focus group topic. For example, if researching cleaning appliances, the set would be made up of pictures from within a domestic environment, and might include people exhibiting different moods – proud, tired, dissatisfied and so on. A typical use for this technique would be to ask the participants which of the people shown is the user of the cleaning appliance under discussion, and then to probe the reasoning behind the responses.

With fixed personality association, the moderator uses a standard set of pictures, irrespective of the topic. Typically these would include a range of different types of people including, for example, young, old, sophisticated, educated and so on. Because the researcher becomes familiar with the personality profile of each image in the set, they can use this as a method for probing the reasons behind the associations people make between a specific photograph and the product or concept being researched. In some respects, this is similar to Product Personality Profiling (see Section 12.4.6). A typical use for fixed association would be to ask participants which of the people shown to them would be likely (or unlikely) to purchase a certain product, or use a certain service and then record the responses on a tally sheet. This gives an overall indication of the perceptions of the group, and also provides a stimulus for further probing.

Situational associations work in a similar way, except that instead of focusing on people, the pictures emphasise specific situations or places. Within the context of human factors/ergonomics and design, this method can be used to explore participants' desires in relation to physical or social environments. For example, pictures of various situations regarding travelling can help people articulate their thoughts about which aspects of the travelling experience they like, dislike or would ideally like to see.

In forced relationship projections, people are asked to make more abstract associations with images from categories such as animals, colours, cars/automobiles or food. Participants are asked to indicate which images most closely relate to the topic being discussed. The choices made provide an indication of the participants' feelings towards the topic and provide interesting perspectives about it. Greenbaum (1998) describes some typical associations:

Animals. Bear (caring, large, friendly). Lion (strong, powerful, not as friendly as bear). Racehorse or greyhound (sleek, streamlined, efficient, sometimes prestigious). Snake, reptile or rodent (normally negative associations, not friendly, unpredictable). Turtle (slow-moving, backwards).

Colours. Hot colours – red, orange, violet (warm, friendly). Cold colours – blue, green (less friendly, distant). Black (distant, mysterious, sleek). White (honest, pure, feminine). Pastel colours (feminine). Deep colours (masculine, sophisticated, expensive).

Food. Burger, steak or potatoes (stable, reliable, consistent). Soups or breads (wholesome, excellent appeal). Vegetables (desirable, good for you). Foreign or exotic food (unusual, not always viewed favourably).

Collages and mood boards (see Section 12.1.7) are similar to these techniques in some respects in that people express their feelings and experiences using images. As with collages, association transcends linguistic limitation. It is an effective technique with adults and children alike. Figure 12.16 shows how association can be used to gain non-verbal feedback from users to inform the design of products, in this case, related to ironing.

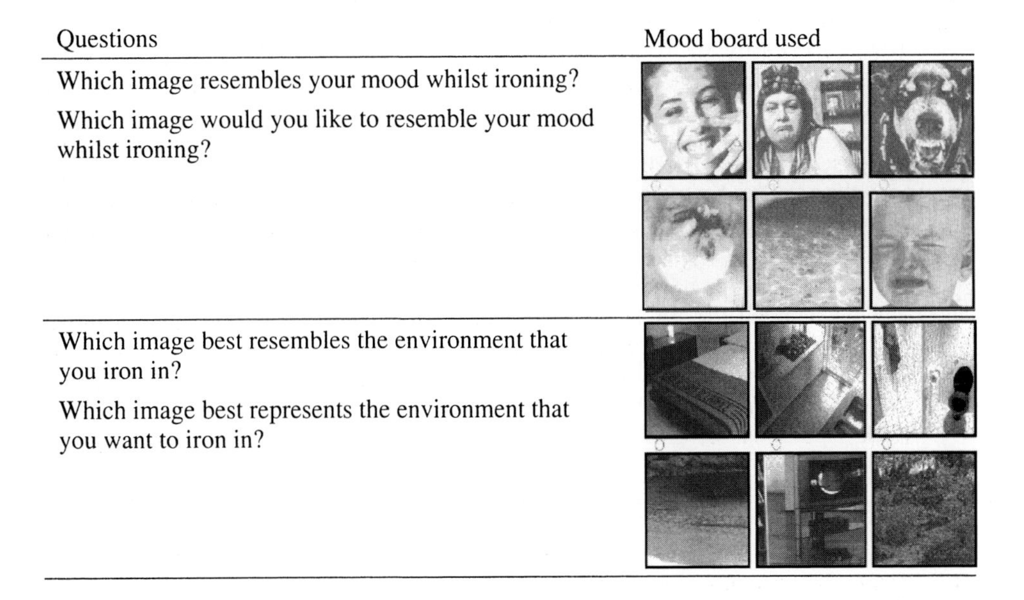

Questions	Mood board used
Which image resembles your mood whilst ironing? Which image would you like to resemble your mood whilst ironing?	
Which image best resembles the environment that you iron in? Which image best represents the environment that you want to iron in?	

Figure 12.16 Mood boards used to explore associations for ironing, ironing products and environments (adapted from McDonagh *et al.*, 2002)

Main advantages:

- provides an insight into peoples' emotional responses to a product or service and helps to reveal the reasons why;
- useful technique for working with children;
- provides a stimulus for discussion and probing.

Main disadvantages:

- requires preparation/availability of suitable images – this can be time-consuming;
- extremely subjective and responses can vary greatly between individuals;
- relies heavily on a skilled interpreter (moderator/researcher).

12.4.8 Conceptual Mapping

Conceptual mapping can be used to explore participants' feelings and perceptions about products by asking them to categorise them, or group them, according to similarity. The ways that people group the products can then be explored in discussion to identify issues that can guide future development.

Greenbaum (1998) describes a simple method. Each participant is given a sheet with a grid on it and allowed five minutes to mark in all the products or items being assessed, grouping them as they think most appropriate (see Figure 12.17). The participants do not have to use the whole grid, and if the grid is too small, they can add more squares. The session can then proceed as follows.

Identify categories and rules used. Discuss the outcomes to find out what categories and rules individuals have used when grouping the items. This can lead to a greater understanding about the rationale participants have used to create the categories, and the types of features they consider when analysing products. At this stage, the actual placement of the specific items is not important, just the categorisation methods used.

Identify predominant approach. Through discussion, establish which categories and rules are predominant. Mark up a labelled grid on a flip-chart to reflect this (moderator).

Achieve consensus. Through discussion, try to achieve consensus as to where each item should be placed. The discussion resulting from this allows greater understanding about the participants' perceptions of each item, why certain items have been placed in certain boxes and the perceived similarities and differences between items.

This technique is relatively simple and provides a framework for further probing. Aiming to achieve consensus (both in terms of categories and item placement) provides a useful focus for the discussion, but the final result is relatively unimportant since it is the information revealed in discussion that is of main interest.

Phone M	Phone A Phone C Phone R	Phone D Phone F
Phone E Phone B	Phone K Phone O Phone P	Phone G Phone H
Phone L Phone N Phone J	Phone I	

Figure 12.17 Example conceptual map showing how a range of telephones might be grouped

Main advantages:

- provides an insight into the ways people analyse products;
- provides feedback on participants' perceptions of specific products or product types;
- easy to set up, minimal pre-preparation;
- provides a stimulus for discussion and probing.

Main disadvantages:

- initial categorisation exercise can be difficult for some participants;
- does not provide definitive 'result'. Not suited to participatory design problems.

12.4.9 Attitudinal Scaling

Attitudinal scaling is a probing technique that is based on identifying the key attributes of a group of products, and exploring how individual products are perceived in relation to these. It provides insights into the features that are considered most important and promotes better understanding of how people assess a product's ability to meet the requirements. It also provides insights into the trade-offs that people make. In the context of human factors/ergonomics and design, an example might be the trade-off between numbers of features (complexity) and ease of use.

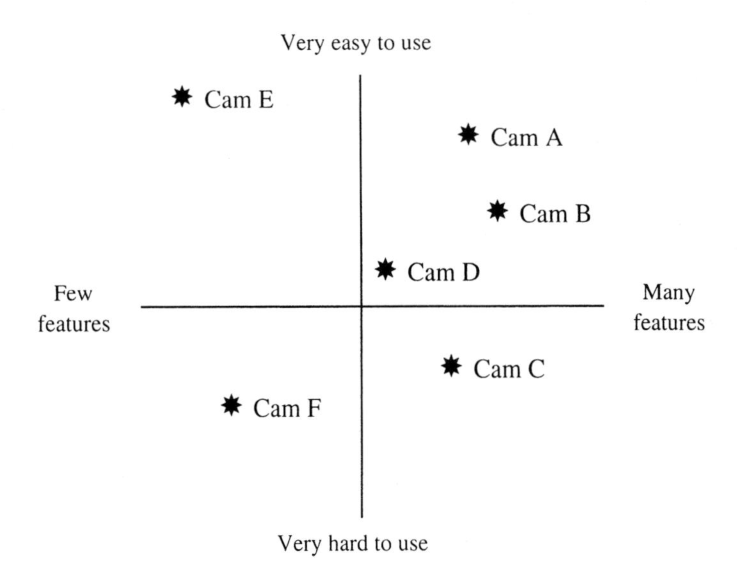

Figure 12.18 Attitudinal scaling exercise showing camcorders as an example

The basic procedure (based on Greenbaum, 1998) is as follows.

Determine key attributes. Discuss the product or topic of interest to identify the most important characteristics, both in terms of positive features/impacts and negative features/impacts. Determine common attributes from these where possible. For example, with camcorders, complexity and ease of use would probably be common themes, but price and reliability might also be important. Decide which attribute to explore further.

Conduct scaling exercise. Give each participant a graph (see Figure 12.18) and ask them to mark on the axes the attributes for exploration, in this case, *complexity* and *ease of use*. Ask each participant to mark on the graph where they think each product lies.

Share outcomes. Draw the graph on a flip-chart so that everyone can see it, and mark up the responses from participants – from just one or two initially.

Probe perceptions. Use the outcomes as a stimulus for probing why people have given their answers and the extent to which others agree or disagree. The responses from other participants can be added to the graph as required. The focus is on trying to understand the participants' perceptions of the products and how and why they make the trade-offs. This can be helpful in revealing and support the understanding of consumer decision-making in purchasing situations.

Main advantages:

- provides an insight into which product characteristics people think are important;
- provides feedback on participants' perceptions of specific products and why they have those perceptions;
- relatively easy to set up – minimal pre-preparation;
- provides a stimulus for discussion and probing.

Main disadvantages:

- only appropriate when trade-offs exist or are relevant to the research;
- scaling exercise can be difficult for some participants;
- some characteristics are hard to assess without hands-on experience;
- does not provide definitive result.

12.4.10 Questionnaire

Questionnaires provide an additional method for capturing feedback from participants during a focus group. They can be used in conjunction with other techniques, for example, visual evaluation, product handling and user testing (Figure 12.19). Questionnaires can be completed early in the session and used as a stimulus. Alternatively, they can be used to summarise participants' views at the end of a session. Another common use in the focus group context is to gather factual information about the participants themselves, although this information is not used in the context of the discussion itself.

There are many different types of questions, including ranking, rating, closed (pre-prepared answers) or open (free response answers). Because the questions are pre-determined, questionnaire results can be more easily analysed than the outputs of a discussion. If there are only a small number of people involved though, the data are unlikely to be statistically secure, yet still valuable nonetheless. Detailed guidance on the design of questionnaires is beyond the scope of this book. For more information, see Sinclair (1995). However, some general principles for their use in the focus group context are given below.

Avoid long or cumbersome questionnaires. The purpose of the focus group is the discussion session, so time spent filling in questionnaires is lost discussion time. Keep the number of questions to a minimum.

Take care with the wording of the questions. They should be easy to understand and unambiguous. Use familiar language and avoid technical terms and acronyms. Keep questions short and to the point.

Avoid leading questions. Do not use questions that might lead the respondent into giving certain answers, for example, 'do you agree that Product A is easier to use than Product B?'.

Design for ease of completion. Limit the number of questions requiring free response answers because these take longer to answer. Closed (tick box) questions, rating scales or ranking are relatively quick to complete.

Use a clear layout. Ensure that the questionnaire is laid out clearly to make it easier to complete and minimise mistakes. If participants are likely to be visually impaired, layout clarity is even more important (see Chapter 7).

Product D **Toaster** **Visual Questionnaire**

1 Regarding the visual appeal of this product, what do you think	very poor 1	poor 2	OK 3	good 4	very good 5	un-sure 0
a) of its **shape and styling**? (e.g. form, proportion, character)						
b) of its **colour**? (e.g. sympathy, suitability)						
c) of the **materials** used? (e.g. texture, strength, surface)						
d) of its **size** and **weight**? (e.g. mass, dimensions, lightness)						

2 What do you **like** about the **appearance** of the product?

3 What do you **dislike** about the **appearance** of the product?

(you can also make comments by attaching notes to the picture)

4 We need to find out whether you have got any previous knowledge of this model.

Please tick in case you

currently own it ☐ use it ☐ used it in the past ☐ have seen it before ☐ have not seen it before ☐

(please tick)

5 Would this product suit your kitchen? yes ☐ no ☐

6 Would you want to buy this product? yes ☐ no ☐

7 Please estimate the retail cost of this product £ ☐

Figure 12.19 Questionnaire used to retrieve aesthetic preferences (from McDonagh *et al.*, 2002)

Main advantages:

- structured format provides a quick way of summarising participants' views;
- standard format means that data can be easily analysed if required;
- provides a good stimulus for discussion and probing if carried out early in session.

Main disadvantages:

- questions are pre-determined, thus limiting the range of responses to some degree;
- pre-preparation required;
- consumes time within the session.

12.4.11 Decision-making Analysis

Used by Caplan (1990) in his focus group studies on photocopiers, decision-making analysis provides numerical results to help understand the participants' views on the relative importance of certain features or functions and the extent to which different designs satisfy these. The outcomes can be helpful, for example, when needing to choose between a number of different design directions.

The analysis is best carried out towards the end of a focus group. This enables the group members to fully understand the designs and allows discussion of key features in the normal way. Sometimes the participants can complete the entire analysis within the session. In some cases, however, the scoring of features may require input from specialists, for example, an engineer's report on technical performance. These will therefore need to be completed at a later date.

Table 12.3 Example decision-making analysis results table (from Caplan, 1990)

	Importance weighting									Goodness rating		
Subject	DC	DS	DT	DB	NB	SM	BC	NH	Avg	A	B	C
Criteria												
Ease of loading scroll	10	10	10	10	9	8	10	10	9.6	10	8	4
Ease of installation/ removal	10	10	10	10	8	7	10	9	9.3	7	10	9
Ease of CFF jam recovery	9	9	8	8	10	6	8	10	8.5	10	7	1
Ease of mainframe jam recovery	9	8	10	9	10	4	8	9	8.4	4	7	10
Ease of platen cover handling	9	9	5	10	6	5	9	10	7.9	8	8	10
Manual copying	4	7	10	7	5	4	7	8	6.5	6	4	10
Visibility of originals	5	4	1	10	7	4	5	7	5.4	4	10	7
Removal of job in mid-scroll	1	6	3	8	9	4	2	8	5.1	10	4	4
Appearance	5	5	7	6	3	6	2	5	4.9	3	10	9
Availability of correctly stapled copies	1	1	1	5	1	1	1	1	1.5	10	1	1
Operability values												
Concept A	447	507	455	596	501	353	452	560	484			
Concept B	499	513	479	604	497	377	471	568	501			
Concept C	444	477	471	559	441	330	440	528	461			

Caplan's technique for decision-making analysis is described below, and an example is shown in Table 12.3.

Establish acceptability criteria. Use the focus group discussion to identify the key criteria for the evaluation. For example, with a photocopier this might be 'ease of jam recovery' or 'appearance'. Identification of these criteria might include some form of brainstorming and list reduction activity.

Rate *importance* of criteria. Each subject rates the importance of each criterion on a scale of 1 to 10, where 10 is the most important. The importance scores or weightings for each subject (identified by their initials) are recorded on the table. In this example, subject DC considers 'ease of loading scroll' to be very important, and 'availability of correctly stapled copies' as unimportant.

Rate the *goodness* of each concept. For each of the acceptability criteria, the group assigns a score reflecting the extent to which each design concept (A, B or C in this case) fulfils the criterion. Again a 1 to 10 scale is used, where 10 is best. In this example, 'ease of scroll loading' is easiest with Concept A, a little less easy with B and much harder with C. For some criteria it may not be possible for the participants to assign a score. For example, if they have no experience of 'jam recovery', then they cannot rate it. In these cases, the scoring will have to be done afterwards by an appropriate expert.

Calculate the *acceptability* values. In this example, acceptability values are *operability* values. The acceptability value is the sum of *importance* multiplied by *goodness* for all the acceptance criteria. Calculate this for each subject for each concept. In this example, subject DC's acceptability value for Concept A (447) is (10×10)+(10×7)+(9×10)+...

If there are clear differences between the concepts, then the preferred option is clear. If the differences are small, closer inspection may be required.

Main advantages:

- requires minimal pre-preparation work;
- provides information about which aspects of the product or system are considered most important by participants;
- produces a tangible output, i.e. an overall view of areas where different concepts perform well or badly and, in some cases, a clear idea of which concept is preferred overall.

Main disadvantages:

- analysis process consumes time during the session;
- it may not be possible to complete the analysis during the session if some aspects require rating by an expert.

12.4.12 Balance Sheets or +/- Charts

Balance sheets are a simple method for analysing ideas or options by considering the advantages (pros) and disadvantages (cons). They help people to evaluate ideas in an organised way and can help groups achieve consensus.

Kettle A	
+	–
Looks good	Heavy
Easy to clean	Handle not very comfortable
Holds a lot	Probably expensive
Great colour	Might not appeal to women
Can see how full it is	

Figure 12.20 Balance sheet example for a kettle design concept

To create a balance sheet, draw a line down a flip-chart to form two columns. Label one with a '+' and the other with a '–'. Ask the group to brainstorm the pros and cons and write these in the appropriate headings (see Figure 12.20). Make sure the sheet is labelled with the idea or product it refers to.

The exercise should be repeated for all the concepts under scrutiny. The completed sheets can then be used to discuss the options and to probe further so that a deeper understanding of key issues can be gained.

Main advantages:

- requires little pre-preparation work;
- focuses people on analysis and helps to clarify thoughts;
- provides stimulus for discussion and probing.

Main disadvantages:

- may be unfamiliar to some participants;
- does not permit numerical analysis.

12.4.13 SWOT Analysis

This is a widely known technique that has many applications in the business world, but is included here because it can help with the evaluation of ideas or concepts. It is fairly detailed and can be time-consuming, so it is best used when there is only a small number of concepts or ideas.

System Design Concept A	
Strengths Very reliable (99.9%) Cheap to maintain Reduced running costs	Weaknesses Not very easy to use High training costs Very expensive to buy Expensive to install
Opportunities Can increase output Can increase quality	Threats Operators may not want to use it

Figure 12.21 SWOT analysis diagram assessing a new production system

SWOT stands for strengths, weaknesses, opportunities and threats. It works by analysing each idea or concept in terms of these headings, commonly using a quadrant diagram as shown in Figure 12.21. In some respects it is similar to the balance sheet method (see Section 12.4.12), but with more refined analysis categories.

It is unsuited to focus groups concerned with consumer products – for these the simpler balance sheet method is more appropriate. It is helpful, however, for groups concerned with the design and/or implementation of more complex systems, for example, for a participatory design group tasked with devising new work processes.

Main advantages:

- requires minimal pre-preparation work;
- focuses on analysis and helps to clarify thoughts;
- useful for participatory design groups.

Main disadvantages:

- can be time-consuming to complete thoroughly;
- unfamiliar method for some people. Not suited to sessions with the general public.

12.4.14 List Reduction

This is a simple filtering technique that helps to reduce brainstormed lists of ideas or concepts down to a manageable size. It is most likely to be used alongside discussion sessions aimed at problem solving, perhaps as part of a participatory design initiative. The steps are as follows.

Clarify ideas. Before attempting any analysis, go through the initial list to ensure everyone understands what the idea is.

Identify filters. As a group, identify the key criteria that must be satisfied if the idea is to remain under consideration. Examples include, cost, likelihood to improve the situation and feasibility. Use of MoSCoW rules (DSDM, 2002) is another example, where requirements are prioritised using the filters *Must have*, *Should have*, *Could have* and *Want to have*.

Review list. Review each item on the list according to the filter criteria. Cross the obvious failures from the list. Mark doubtful cases in brackets.

Repeat process. Carry out the process again if required, focusing on the remaining items, and possibly defining new filter criteria.

Main advantages:

- requires minimal pre-preparation work;
- focuses on analysis and helps to clarify thoughts;
- reduces large lists to a manageable size.

Main disadvantages:

- some participants may be unfamiliar with the method;
- requires a skilled moderator to ensure consistency and objectivity.

12.4.15 Voting and Ranking

Voting and ranking are ways of gauging the relative popularity of the products, ideas or concepts being discussed, and to help make choices. The outcomes can be used to help stimulate further discussion or probing by the moderator. Participants often like to carry out voting or ranking exercises because they provide some form of tangible result or outcome for the session.

There are several ways of setting up voting and ranking schemes, but the key consideration for their use within group discussions is speed and ease of use. There is little point in using methods that require complex statistical analysis and provide results that are relatively meaningless to the participants. Also, due to the relatively small number of people taking part, the reliability of the outputs is limited. It is important to recognise the limitations of voting and ranking in this context, and to use the outputs appropriately. A few simple voting and ranking schemes are described below. This is not intended as an exhaustive list. The reader should adapt these methods to suit their own particular needs.

Pair comparison. The options are compared against each other in pairs, i.e. Option 1 versus 2, then 1 versus 3 and so on. Participants carry out the exercise individually, recording their results on a simple grid. The participants are asked to identify which of each pair is better. In the example shown (Figure 12.22), Chair 1 has been judged better than Chairs 2, 4 and 5, but worse than Chairs 3, 6, 7 and 8.

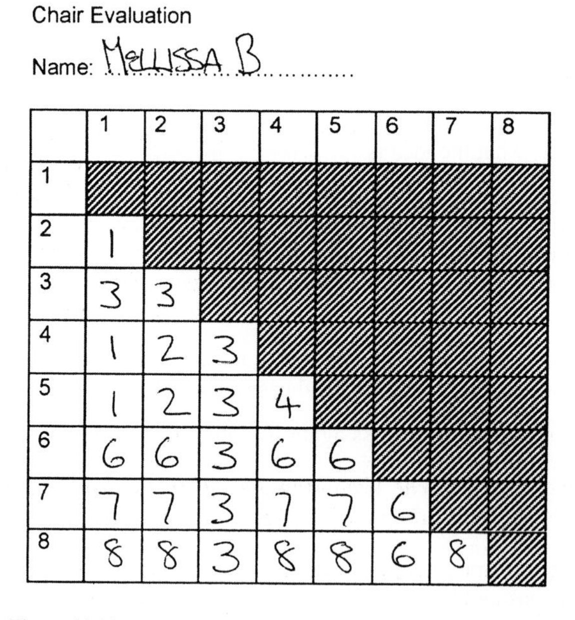

Chair Evaluation

Name: MELISSA B

	1	2	3	4	5	6	7	8
1								
2	1							
3	3	3						
4	1	2	3					
5	1	2	3	4				
6	6	6	3	6	6			
7	7	7	3	7	7	6		
8	8	8	3	8	8	6	8	

Figure 12.22 A grid used for pair comparison of eight chair designs

As an alternative, it can sometimes be helpful to allow pairs of options to be judged as equal. When completed, the results are tabulated and scores are assigned using a simple system, such as one point for each time an option is judged better than another.

Dot voting. The choices on offer are written on flip-charts or display boards, and participants vote by placing stick-on coloured dots against their choices as appropriate. Participants may be given one dot or more according to need.

Ballot voting. Participants vote for one or more of the options by use of secret ballot. Votes are recorded on cards and handed in to the moderator.

Ranking. Each participant ranks the options in order of preference. If there are a large number of options, the process may be simplified by limiting the ranking to the top five or three choices. Ranking scores are determined using a simple system, for example, five points each time an idea is ranked best, four points for second place, and so on. The results can be called out by each participant and tabulated by the moderator, or the process can be carried out secretly, as in the nominal group technique (see Section 12.3.2).

Main advantages:

- basic voting and ranking methods are simple and adaptable;
- voting and ranking helps to prioritise options;
- provides a tangible output.

Main disadvantages:

- can be time-consuming to complete and to analyse;
- unlikely to have statistical validity.

12.5 CONCLUSIONS

This chapter brings together a range of tools and techniques that can be used to extend the usefulness of focus group methods. By using these tools in conjunction with the practical guidance on logistics given in Chapter 2, designers and researchers in human factors/ergonomics can exploit the full potential of focus groups.

Flexibility is a key strength of the focus group method and the same is true of the tools included in this chapter. The descriptions given here are not prescriptive – researchers are encouraged to experiment, adapting the techniques as appropriate to suit their own requirements.

12.6 REFERENCES

Buzan, T. and Buzan, B., 1993, *The Mind Map Book*, (London: BBC Books).

Caplan, S., 1990, Using focus group methodology for ergonomic design. *Ergonomics*, **33**, pp. 527-533.

Channel 4, 2000, *Better by Design*, broadcast July/August 2000. Video available from BSS, Better by Design, PO Box 1001, Manchester, M60 3JB, UK. See also www.design-council.org.uk/betterbydesign/ (accessed 30 April 2002).

Clegg, B. and Birch, B., 1999, *Instant Creativity*, (London: Kogan Page Ltd).

DSDM, 2002, Dynamic Systems Development Method, www.dsdm.org (accessed 1 March 2002).

Dolan, W.R. Jr., Wiklund, M.E., Logan, R.J. and Augaitis, S., 1995, Participatory design shapes future of telephone handsets. In *Designing for the Global Village*, Proceedings of the Human Factors and Ergonomics Society 39th Annual Meeting. Volume 1, 9–13

Oct 1995, San Diego, California, (Santa Monica, CA: Human Factors and Ergonomics Society), pp. 331-335.

Ede, M., 1999, Focus Groups to Study Work Practice, www.useit.com/paper/meghan_focusgroups.html (accessed 13 December 1999).

Greenbaum, T.L., 1998, *The Handbook for Focus Group Research*, (Thousand Oaks: Sage Publications).

Higgins, J.M., 1994, *101 Creative Problem Solving Techniques*, (Florida: New Management Publishing Company).

Kuhn, K., 2000, Problems and benefits of requirements gathering with focus groups: a case study. *International journal of human-computer interaction*, **12** (3 & 4), pp. 309-325.

Kuorinka, I., 1997, Tools and means of implementing participatory ergonomics. *International Journal of Industrial Ergonomics*, **19**, pp. 267-270.

McCallion, S., 1999, Experience the Future: Behaviour Anticipation in Product Development. *Innovation*, Winter 1999, (Dulles, VA: Industrial Designers' Society of America), pp. 52-55.

McClelland, I., 1995, Product assessment and user trials. In *Evaluation of Human Work*, Wilson, J.R. and Corlett, E.N. (eds), (London: Taylor and Francis), pp. 249-284.

McDonagh, D., Bruseberg, A. and Haslam, C., 2002, Visual product evaluation: Exploring users' emotional relationships with products. *Applied Ergonomics*, **33** (3), pp. 237-246.

McNeese, M.D., Zaff, B.S., Citera, M., Brown, C.E. and Whitaker, R., 1995, AKADAM: Eliciting user knowledge to support participatory ergonomics. *International Journal of Industrial Ergonomics, Special Issue: 'Participatory Ergonomics'*, **15**, pp. 345-365.

Nielsen J., 1997, Usability Testing. In *Handbook of Human Factors and Ergonomics* (2nd edition), Salvendy, G. (ed.), (New York: John Wiley and Sons), pp. 1543-68.

O'Brien, D.D., 1981, Designing systems for new users. *Design Studies*, **2**, pp. 139-150.

Rainbird, G. and Langford, J., 1998, Feasibility study of containerisation for travelling post office operations. In *Contemporary Ergonomics 1998*, Hanson M.A. (ed.), (London: Taylor and Francis), pp. 381-385.

Sato, S. and Salvador, T., 1999, Playacting and focus troupes: theatre techniques for creating quick, intense, immersive and engaging focus group sessions. *Interactions. New Visions Human-Computer Interaction*, **6** (5), pp. 35-41.

Sinclair, M.A., 1995, Subjective assessment. In *Evaluation of Human Work*, Wilson, J.R. and Corlett, E.N. (eds), (London: Taylor and Francis).

Snow, M.P., Kies, J.K., Neale, D.C. and Williges, R.C., 1996, Participatory Design. *Ergonomics in Design*, **4** (2), pp. 18-24.

Wilson, J.R., 1991, Design Decision Groups – A participative process for developing workplaces. In *Participatory Ergonomics*, Noro, K. and Imada, A. (eds), (London: Taylor and Francis). Also in Wilson, J.R., 1995, Ergonomics and participation. In *Evaluation of Human Work*, Wilson, J.R. and Corlett, E.N. (eds), (London: Taylor and Francis), pp. 1071-1096.

Author Index

Index